Read this!

Read this!

Business writing that works

ROBERT GENTLE

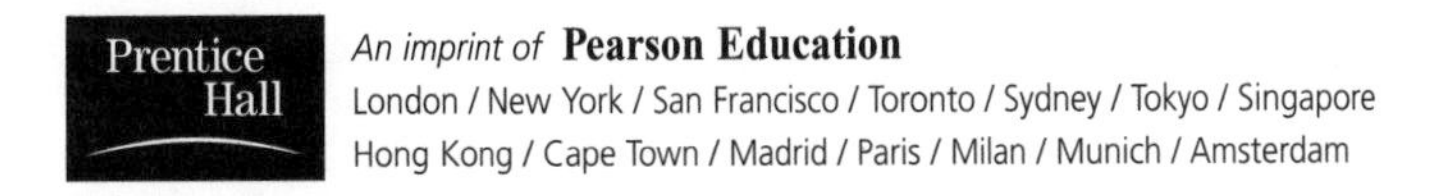

Prentice Hall

An imprint of **Pearson Education**

London / New York / San Francisco / Toronto / Sydney / Tokyo / Singapore
Hong Kong / Cape Town / Madrid / Paris / Milan / Munich / Amsterdam

PEARSON EDUCATION LIMITED

Head Office:
Edinburgh Gate
Harlow CM20 2JE
Tel: +44 (0)1279 623623
Fax: +44 (0)1279 431059

London Office:
128 Long Acre, London WC2E 9AN
Tel: +44 (0)20 7447 2000
Fax: +44 (0)20 7240 5771
Website:www.business-minds.com

First published in Great Britain in 2002

ISBN 0 273 65650 3

British Library Cataloguing in Publication Data
A CIP catalogue record for this book can be obtained from the British Library.

10 9 8 7 6 5 4 3 2 1

Typeset by Pantek Arts Ltd, Maidstone, Kent.
Printed and bound in Great Britain by Biddles Ltd, Guildford & King's Lynn.

The Publishers' policy is to use paper manufactured from sustainable forests.

About the author

Robert Gentle, aged 44, has a degree in mechanical engineering and did post-graduate studies in aeronautical engineering and computer science.

After two years working as an engineer and computer analyst/programmer, he embraced his first love – writing – and became a financial journalist.

Over the next 15 years, he wrote for numerous business publications, won a number of journalism awards, travelled extensively, moved into corporate public relations and eventually became a sought-after expert on good business writing. He has conducted over 100 writing workshops and spoken at several international conferences.

"Poor documents are so commonplace that deciphering bad writing and bad visual design have become part of the coping skills needed to navigate in the so-called information age."

KAREN SCHRIVER
Dynamics in Document Design

Towards a new world standard

When was the last time you read a business document and understood it on the first pass? Chances are you had to re-read it carefully – assuming, of course, you didn't just throw it away in frustration.

As we all know from personal experience, most business writing is indigestible. The main reason is that there are no globally agreed standards of clarity and readability: everyone's doing their own thing, and the result is a hodge-podge of different styles, layouts and structures.

We could take a leaf out of the book of accountants who, to their credit, have established some degree of consistency in the way financial statements are drawn up around the world. The profession's international guidelines are known as GAAP, or *Generally Accepted Accounting Practice*.

In this book, we're going to establish global standards of clarity and readability that can easily be applied to all kinds of business documents – from letters and reports to e-mail and prospectuses.

Who knows: maybe they will eventually come to be known as GAWP – *Generally Accepted Writing Practice*. Remember – you read it here first.

Robert Gentle

"For more than 40 years, I've studied the documents public companies file. Too often, I've been unable to decipher just what is being said or, worse yet, had to conclude that nothing was being said."

WARREN BUFFETT
Chairman of Berkshire Hathaway

Contents – overview

CHAPTER 1 How to grab your reader's attention

Theory

Practice

Real-world case studies

CHAPTER 2 How to create a user-friendly layout

Theory

Practice

Real-world case studies

CHAPTER 3 Letters, faxes and e-mail

Theory

Practice

Real-world case studies

CHAPTER 4

Reports

Theory

Practice

Real-world case studies

"When British Gas sent out its first dividend statement to shareholders, it was so unclear that 10 000 thought it was a bill and sent a cheque. It is just not good enough."

INDEPENDENT (UK)
22 October 1997

PART 1 The basics

"To do our work, we all have to read a mass of papers. Nearly all of them are far too long. This wastes time, while energy has to be spent in looking for the essential points."

WINSTON CHURCHILL

in a memo to his staff in 1940 on the importance of brevity

CHAPTER 1 How to grab your reader's attention

Clearing up some common myths

***Myth:* you must use formal language**
***Reality:* clear, everyday language is better**

Words are a means to an end, not an end in themselves. They can make you cry in a sentimental novel, rouse you to action in a stirring speech or, when used in a memo, send you running to an important meeting. Most business writing is in the latter category – it is functional. Unlike literature, business writing is not read for pleasure. It is written to be understood and acted on; afterwards, it is filed or thrown away. So use plain, everyday words that get your point across.

***Myth:* readers have to be fully informed**
***Reality:* readers only have to be reasonably informed**

The purpose of business writing is to persuade – eg, buy my product; no, we won't pay your insurance claim; Microsoft is a *buy*. Therefore, you only need to give readers enough information to enable them to make a reasonably informed decision. Long, detailed writing that tries to cover all bases is self-defeating. As Mark Twain once said: few sinners are saved after the first 20 minutes of a sermon. It's the old 80:20 rule again: 20% of what you say accounts for 80% of your impact.

***Myth*: people are dying to read your document**
***Reality*: they would rather be doing something else**

Let's face it: we think that when the document we've written lands on the reader's desk, he's going to stop his work, ask his secretary to hold all calls, put his feet up and read it with unswerving interest. In reality, he has no time (information overload) and even less interest (he probably doesn't even know you). Unlike a favourite book or magazine, which we *choose* to read, business writing is something we *have* to read. We don't pay for it. It interrupts our work. That's why we file it or throw it away if it doesn't make its point quickly.

"More traditional companies are beset with a different form of denial: the wish-fulfilment fantasy that people [on the Internet] cannot resist their Chairperson's letter to stockholders from the current annual report. Of course, we know the truth: we came, we clicked, we yawned."

CHRISTOPHER LOCKE
Microsoft Internet Magazine, 4 August 1997

What your reader *really* thinks

His most common reaction is – do I really have to read this?

Document	*Average reading time*	*Your attitude*
Book or novel	3–8 hrs	You invest time and energy. The subject matter interests you and you've paid for it.
Lifestyle magazine	1–2 hrs	Your leisure subjects, from cars to wine. You often subscribe to these magazines and enjoy reading them.
Current affairs magazine or newspaper	10 mins – 1 hr	This helps you keep up with what's happening in your industry sector. You often stop reading it when you're on vacation.
Report (general)	3–8 mins	If you get past the executive summary, you'll read until your interest flags.
Annual report	5–10 mins	You'll skim through it and pick up the points that interest you.
Website	1–5 mins	If you don't find anything interesting, you're "outta there".
Business correspondence	$\frac{1}{2}$–2 mins	If you don't get the point, you query it, file it or throw it away.
Memo	$\frac{1}{2}$–1 min	If you don't get the point, you query it, file it or throw it away.
Ad, flyer, brochure	5–30 secs	If it doesn't press your hot buttons, you throw it away.

"To be clear is to be efficient; to be obscure is to be inefficient. Your style ... is to be judged not by literary conventions or grammatical niceties, but by whether it carries out efficiently the job you are paid to do."

SIR ERNEST GOWERS
The Complete Plain Words

Five steps to grabbing your reader's attention

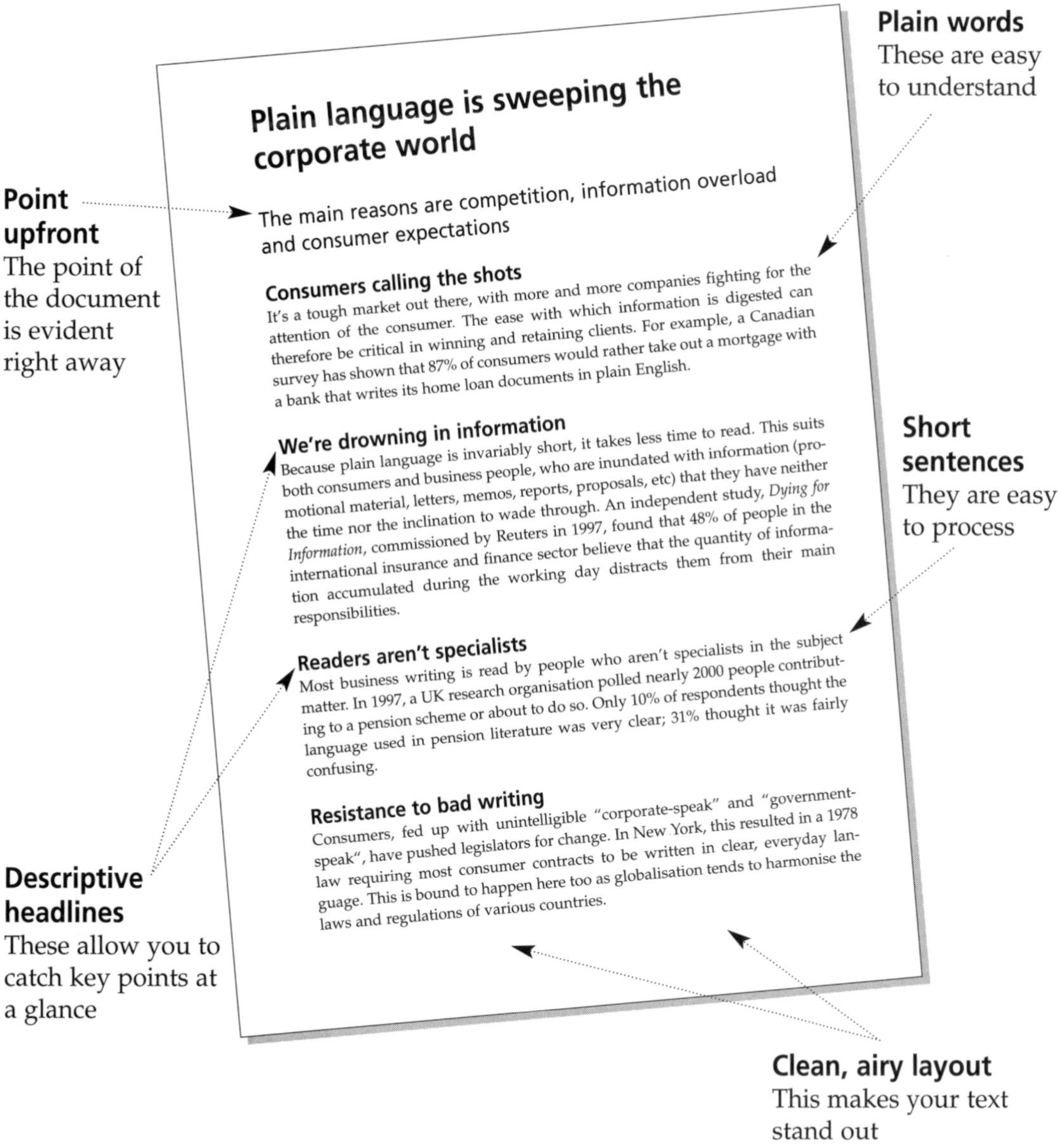

Plain language is sweeping the corporate world
The main reasons are competition, information overload and consumer expectations
Consumers calling the shots
It's a tough market out there, with more and more companies fighting for the attention of the consumer. The ease with which information is digested can therefore be critical in winning and retaining clients. For example, a Canadian survey has shown that 87% of consumers would rather take out a mortgage with a bank that writes its home loan documents in plain English.
We're drowning in information
Because plain language is invariably short, it takes less time to read. This suits both consumers and business people, who are inundated with information (promotional material, letters, memos, reports, proposals, etc) that they have neither the time nor the inclination to wade through. An independent study, *Dying for Information*, commissioned by Reuters in 1997, found that 48% of people in the international insurance and finance sector believe that the quantity of information accumulated during the working day distracts them from their main responsibilities.
Readers aren't specialists
Most business writing is read by people who aren't specialists in the subject matter. In 1997, a UK research organisation polled nearly 2000 people contributing to a pension scheme or about to do so. Only 10% of respondents thought the language used in pension literature was very clear; 31% thought it was fairly confusing.
Resistance to bad writing
Consumers, fed up with unintelligible "corporate-speak" and "government-speak", have pushed legislators for change. In New York, this resulted in a 1978 law requiring most consumer contracts to be written in clear, everyday language. This is bound to happen here too as globalisation tends to harmonise the laws and regulations of various countries.
Plain words
These are easy to understand
Point upfront
The point of the document is evident right away
Short sentences
They are easy to process
Descriptive headlines
These allow you to catch key points at a glance
Clean, airy layout
This makes your text stand out

"If you can't write your idea on the back of my calling card, you don't have a clear idea."

DAVID BELASCO
Theatre producer

Make your point upfront

Don't bury your message in some distant paragraph your reader is unlikely to reach

Big picture first, detail afterwards

The mind looks for broad meaning first, then detail. Your writing should therefore first make sense from a big picture perspective – ie, it should contain some kind of upfront summary. Obviously, you'll need to organise your thoughts first; if the summary isn't clear in your head, it won't be clear in your document.

Summary shows general direction

Reading a document without an upfront summary is like going into a meeting without an agenda – you've got no sense of your final destination. You have to read the whole document to find its meaning. So you're likely to throw it back into your in-tray – or even more likely, into the bin.

Conclusion at the front, not the end

At school, we were taught to structure a document in three parts: Introduction, Body and Conclusion. That doesn't work in the real world where most people don't have the time to go on a treasure hunt for meaning. State your conclusion upfront, and use the rest of the document to support and explain that conclusion.

What should the summary contain?

A descriptive headline is a must, especially for titles and chapter headings. If a section in your report is on how badly maintained roads cause car accidents, call it: *How badly maintained roads cause car accidents*. Where appropriate, your headline may be followed by a 10- to 20-word summary (like the one at the top of this page).

"One day [FBI chief J. Edgar Hoover] received [a memo] whose margins were too small. In big, red letters he scrawled an angry warning across the top: 'Watch the borders!' The next morning, his frightened assistants transferred 200 agents to Canada and Mexico."

JULIA VITULLO-MARTIN and J. ROBERT MOSKIN
Executive's Book of Quotations

Use a clean, airy layout

It is attractive, user-friendly and makes your text stand out

Good layout pulls the reader in

Layout is the first thing you notice, even before a single word is read. If the layout is clean and airy, it pulls you into the document. Bad layout, characterised by dense, unreadable type stretching from one side of the page to the other, puts you off as nothing else can. It's about as inviting as a slap in the face.

Use lots of white space

Use a narrow column of text, because this maximises your available white space – which in turn makes your writing stand out. Even if you use two columns (newspaper-style), keep some white space running down the side. Don't ever fill your page with text in an effort to save paper. All that achieves is an unreadable page – and *that's* a waste of paper.

Choose the right font

Try to use serif fonts in the main body of your text. (Serifs are those little connecting strokes at the beginning and end of each letter; they make words easier to read.) Examples are Book Antiqua and Times New Roman. Sans serif fonts (sans is French for "without") often work well in headlines, as you can see from this book. Examples are Arial and Helvetica.

Use capitals sparingly

Have you ever seen a newspaper written in capitals? Of course not. Capitals are hard to read because they consist of harsh, straight lines bunched close together. Use capitals sparingly – for example, in highlighting titles, headings and names. Where appropriate, mix and match with other fonts. Example: The HARRY POTTER books, by *JK Rowling*, are popular with kids.

See page 41 for more on good layout

"Headings are … a technique that forces good organisation. At the very least, headings require people to arrange their documents into blocks of information instead of scattering ideas throughout."

EDWARD P. BAILEY, JR
The Plain English Approach to Business Writing

Use descriptive headlines

They pull you through the document, allowing you to catch key points at a glance

Text without headlines is a pain!

Don't you just hate it when you're faced with a page of mind-numbing, wall-to-wall ink without a headline in sight? That's because it forces us to wade through the text and mentally break it up into digestible chunks. Often, we do this with those coloured highlighter pens. Save your reader the hassle: break the text up yourself and flag each section with a headline.

Make your headlines descriptive

A descriptive headline describes what's about to follow. In an economic report, *Inflation heading up again* is a descriptive headline. *Inflation* isn't – all it does is beg the question: what about it? Similarly, in a promotional brochure, *We can reduce your tax* is a descriptive headline; *Tax* isn't. For good examples of descriptive headlines, look no further than your daily newspaper.

Introduction and other empty headlines

All *Introduction* says is that you're at the start of a section – as if you didn't know that already. If your introduction explains why the document was written, rather call it *Why this document was written*. Other empty headlines include *Prospects, Performance, Market Share*, etc. Make these come alive. For example, *Prospects – a good year ahead* is better than just *Prospects*.

Use headlines in graphs too

Headlines can make a graph come alive and pull the reader right in. For example, if your graph shows spiralling costs in your department, you may want to slap a bold headline *Going through the roof* across the top.

"Because English is so rich and versatile, you can usually say what you mean in short, vigorous, everyday words which most of your readers will find familiar."

THE PLAIN ENGLISH CAMPAIGN (UK)
The Plain English Story

Use plain, everyday words

No one has ever complained that something is too easy to understand

Simple is best

Try, don't endeavour; use, don't utilise; get into your car, not your vehicle. Simple words are easily understood and have greater impact than longer words. Remember that business writing has a very short shelf-life; it will be in the bin in no time. Save your literary prowess for something designed to elicit emotion, like a speech or a personal letter.

Long	*Short*
manufacture	make
purchase	buy
vendor	seller
terminate	end
provenance	origin
parentheses	brackets
expeditiously	quickly

Avoid redundant qualifiers

Surprise means something that is unexpected; an unexpected surprise is therefore meaningless. To join means to bring together; to join together is meaningless. William Brohaugh, author of *Write Tight*, recommends you weed out redundant qualifiers by asking: "As opposed to what?" Example: green in colour. As opposed to what – green in height?

Redundant	*Plain*
foreign imports	imports
joint agreement	agreement
major breakthrough	breakthrough
disappear from sight	disappear
consensus of opinion	consensus
advance notice	notice
future prospects	prospects

"If we make a habit of saying 'The true facts are these', we shall come under suspicion when we profess to tell merely 'the facts'. If a crisis *is always acute, and an* emergency *always grave, what is left for these words to do by themselves? An* unfilled vacancy *may leave us wondering whether a mere vacancy is really vacant."*

SIR ERNEST GOWERS
The Complete Plain Words

Steer clear of jargon

These days, people aren't fired; they are vocationally relocated. The economy isn't slowing down; it's experiencing negative growth. A spacecraft toilet is a post-nutritive disposal unit. Every industry has its jargon. So long as industry experts are speaking to other industry experts, the use of jargon is not a problem. However, most business writing is between experts and lay people (customers, suppliers, legislators – even experts in other fields). So use straight talk.

Jargon	*Straight talk*
pre-owned vehicle	used car
anti-personnel projectile	bullet
localised capacity deficiency	bottleneck
inter-dental stimulator	toothpick
negative patient care outcome	death
geographically relocate	transfer
rodent operative	rat catcher
utility access hole	manhole

Avoid longwinded expressions

Longwinded expressions are becoming so common that we no longer even question them. *Now* has become *at this point in time. If* has become *in the event of.* Here's a classic: *Enclosed herein is your tax return in respect of the year 2002.* Translation: *Here is your tax return for 2002.*

Longwinded	*Short*
in respect of	for, about
having regard to	about, for
in the majority of instances	usually
as a consequence of	because
in excess of	more than
in the course of	during
for the purpose of	to
at the present time	now
prior to	before
subsequent to	after

"Vigorous sentence writing is concise. A sentence should contain no unnecessary words … for the same reason that a drawing should have no unnecessary lines and a machine no unnecessary parts."

WILLIAM STRUNK and E.B. WHITE
The Elements of Style

Use short sentences

They get your point across quickly and persuasively

Use active voice more than passive

Active sentences use personal pronouns (eg, I, we, they) and are therefore short and punchy. Example: *We've located the book you ordered and will try to get it to you by tomorrow.* Passive sentences, while at times useful and necessary, are nevertheless weak and limp. Example: *The book you ordered has been located and every endeavour will be made to get it to you by tomorrow.* Readers relate to active sentences; use them as often as you can.

Avoid redundant words

Redundant words creep into sentences unnoticed. Take them out. For example, watch how the following sentence shrinks:

Long:	The information that we currently have at our disposal
Better:	The information we have
Even better:	Our information (depending on the source)

Don't pad the start of your sentence

We often start sentences with *I don't have to tell you* or *For your information*. These empty remarks state the obvious. If something is interesting, why preface your sentence with *Interestingly*? And if it's not interesting, why use it? Your sentence doesn't need a crutch: let it stand on its own two feet.

Dust off that grammar book

Spell-checkers, grammar-checkers and sentence analysis software are fine in the quest for tight sentences, but it's your grasp of grammar and vocabulary that's the deciding factor. It's what enables you to cut the fat from a long, unwieldy sentence and leave the muscle.

"Executives at every level are prisoners of the notion that a simple style reflects a simple mind. Actually, a simple style is the result of hard work and hard thinking."

WILLIAM K. ZINSSER
On Writing Well

Long sentences made short

1 **In the event that one or more pages of this fax message is missing or difficult to read, you are requested to call the following number: (011) 555-1234.**

If any of these pages are missing or difficult to read, please call (011) 555-1234.

2 **Customers find that using an ATM for banking transactions is not only more convenient, but also cheaper than dealing with tellers across the counter. They find this latter type of transaction time-consuming.**

Customers find ATM transactions more convenient, cheaper and less time-consuming than dealing with tellers.

3 **In the event of a price increase, we will renegotiate the contract that we currently have with you.**

If prices increase, we will renegotiate our contract.

4 **The research that we currently have at our disposal suggests that Company A will acquire the entire issued share capital of Company B for a total consideration likely to be in the vicinity of $15m.**

The research we have suggests that Company A will acquire Company B for about $15m.

5 **There are many legitimate tax deductions which clients can avail themselves of. These deductions, which operate through Section 28 of the Income Tax Act, are usually only available to short-term insurance companies. They include:**

Section 28 of the Income Tax Act allows clients the following legitimate tax deductions, which are usually only available to short-term insurers:

"Ingest, imbibe, and be full of cheer, because the next day you and I shall bite the dust.

(Eat, drink and be merry, for tomorrow we die.)"

PETER GORDON
Verbiage for the Verbose

6 **The braking mechanism of the new BMW belonging to the Marketing Director was found to be operationally defective, which necessitated the return of the vehicle to the dealership.**

The Marketing Director's new BMW was returned to the dealership because the brakes weren't working properly.

7 **Consideration is being given to the acquisition by our group of a stake in Xylon Tech Corporation. The rationale for such an initiative would be to augment the group's skills base in the area of technology and to enhance its capacity to generate more sales revenue in those markets represented by Asia and Australia.**

We are thinking of acquiring a stake in Xylon Tech Corporation in order to increase our technology skills and generate more business in Asia and Australia.

8 **In the event of a tenant finding that a committee room booked for a meeting is not available, the tenant shall contact the staff member on duty at reception who will then allocate that tenant an alternative venue.**

If the committee room that you have booked is not available, please inform reception and they will provide you with an alternative one.

9 **You shall give us notice within seven days of any material change affecting the risk attaching to any insured property. We may at our own option, after such notice, elect to discontinue the insurance or alternatively to continue the insurance subject to an increased premium.**

You have seven days in which to tell us of any material change that may increase the risk of loss or damage to your property. We will then have the right either to end the policy, or let it continue by asking you to pay an increased premium.

"Suppose the marketing department is asked to assess whether a new product will make a profit. At a traditional company, it would issue a 100-page report that would include market surveys, demographic assumptions, economic scenarios ... and more.

If you really want someone to evaluate a project's chances, only give them a single page to do it – and make them write a headline that gets to the point, as in a newspaper. There's no mistaking the conclusion of a memo that begins: 'New Toaster Will Sell 20 000 Units for $2m profit'."

RICARDO SEMLER
Maverick

Real-world case studies

The following case studies are based on real documents from various companies and organisations around the world.

The original versions have been rewritten using the techniques you've just learned.

1 Letter
2 E-mail
3 Powerpoint slide
4 Website page
5 Public sign
6 Newsletter article
7 Graphs

It should be obvious by now that good writing is good writing wherever you may find it. Principles of clarity and readability are universal, and apply in any medium where words and sentences are strung together.

1. Letter to the manager of a major food store

Dear Sir

I've been meaning to write this letter to you for a while now about a problem that I – and no doubt many other shoppers – continually have at your store.

It happened again last week – hence this letter. I am, of course, referring to the fact that whenever one shops at your store from around 3.30pm onwards, it is painstakingly obvious that supplies of fresh produce on your shelves start to run dangerously low.

By around 5pm, there is little or nothing left – no fresh juices, no cheeses, no yoghurt, no meat, no lettuce, etc. Not only is this highly irritating for late-night shoppers; it is also somewhat bizarre because your press, TV and billboard advertising campaign constantly makes reference to the fact that you stay open late – sometimes until 9pm. It seems to me that there is little point in staying open late if you cannot keep your shelves reasonably stocked. Surely, with today's computer technology and basic stock-control procedures, you should be able to manage this quite easily?

Sincerely

What's wrong?

1. No headline to set the tone.
2. Too long.
3. Point buried in the second paragraph.

Dear Sir

So much for late-night shopping

When are you going to ensure that cheese, lettuce and other fresh produce doesn't run out towards mid-afternoon?

You advertise that your store remains open as late as 9pm – but you don't keep your produce shelves reasonably stocked until 9pm. What's the point?

Surely with computer technology and basic stock control procedures, you should be able to ensure that these shelves are adequately stocked until closing time?

Sincerely

2. An e-mail aimed at selling advertising in a magazine

SUBJECT *Comm-Direct* – the magazine for today's communicators

Dear Susan

As a busy executive responsible for marketing your company's services, you're probably flooded with requests for advertising. Which means you become selective about which publications you choose to advertise in. It's a question of successful adspend – of getting your marketing message to the largest reach – in the right target market – for the most efficient advertising pound.

Comm-Direct is a monthly magazine that reaches the cream of communications professionals – marketing managers throughout the country as well as public relations practitioners, both corporate and independent PR consultants and advertising agencies. It is a publication that reaches precisely the people who have a budget for your services. A budget that will interest you.

When we look at our readership research, some interesting facts emerge. Facts that may surprise you. Like the fact that nearly 70% of our readers are working in large corporates, and only 29% in consultancies or small offices; that our readers are at a very senior level with the majority holding director or manager designations in communications, PR, marketing, advertising or corporate affairs, and 8% being CEOs or MDs; that 62% are real decision-makers and take the final decision in respect of their job responsibilities; and 55% are earning salaries in the A-income bracket.

Comm-Direct therefore makes more sense. And it certainly makes business sense for you to advertise to this corporate market. *Comm-Direct* is perfectly targeted to a market of communications managers who are captive, empowered and receptive to this advertising. With a full-page, full-color advert for only £1 000, or a quarter-page, black-and-white advert for only £550, you can reach the 20 000 top communications executives interested in your message.

I would value the opportunity to meet with you and review this in more detail.

Sincerely

What's wrong?

1 Meaningless headline.
2 Long and convoluted.
3 Statistical overkill in third paragraph – better to focus on one or two key figures.

SUBJECT Reach 20 000 communications professionals each month

Dear Susan

You'll reach the cream of Britain's communications professionals if you advertise in the monthly magazine *Comm-Direct*.

Most of our 20 000 readers are directors or managers in communications, PR, marketing, advertising and corporate affairs. Over 70% of them work in large corporates – and that's where the money is for your services.

A full-page colour ad costs only £1 000 and a quarter-page black/white ad £550.

If you are interested, may I call you for a quick meeting to discuss this in more detail?

Sincerely

3. A slide from a Powerpoint presentation on export strategy

Proposed plan to get our export drive into Africa under way

- Corporate headquarters to be established and located in Johannesburg (latest date: January)
- Pending joint-venture agreement in respect of warehousing to be finalised (latest date: February)
- Delivery of first batches of Personal Computer units from our UK plant to commence once warehousing in place (latest date: March)

What's wrong?

1 Long sentences that are difficult to catch at a glance.
2 Language used not plain enough.
3 Excessive use of passive voice.

How we get started in Africa

- **January** Set up HQ in Johannesburg
- **February** Finalise warehousing joint-venture
- **March** Start shipping in PCs from our UK plant

4. From a corporate insurer's website

Manning & Rorsch in comparison to a captive insurance company

The advantages of companies utilising a Manning & Rorsch self-insurance facility over a dedicated captive insurance company of their own include:

- The licence granted to individual applicants for captive insurance licences is usually restricted to a particular class of business, whilst Manning & Rorsch is licensed to transact all six classes of business. Manning & Rorsch is specifically designed for "sectional title" captive insurance company ownership, and is fully approved for this purpose by the Registrar of Insurance.
- Economies of scale are achievable in Manning & Rorsch through the supply of the specialist resources, administration and systems required by the Financial Services Board for the running of a captive insurance company.
- All the management responsibilities associated with a wholly-owned captive insurance company are assumed by Manning & Rorsch, who already employ the necessary expertise.
- A minimum capitalisation of £50m would be required by an individual applicant to establish a captive insurance company (subject to a minimum asset base of £30m), compared to a minimum solvency margin required in a Manning & Rorsch cell of only 25% of premium income.
- The owners of a wholly owned captive insurance company are obliged to consolidate the captive; as the shareholding of a Manning & Rorsch participant in Manning & Rorsch is small, the participant is not obliged to consolidate its investment in Manning & Rorsch.

What's wrong?

1. Nothing is obvious at a glance.
2. Too much narrative and not enough headlines.
3. Long words and complex sentences.

Why Manning & Rorsch is better than a wholly owned captive

Fully approved

Manning & Rorsch is specifically designed for "sectional title" captive ownership and is fully approved for this purpose by the Registrar of Insurance.

An all-embracing licence

Manning & Rorsch is licensed to transact all six classes of business while a dedicated or wholly owned captive is usually restricted to a particular class of business.

Economies of scale

Manning & Rorsch provides all the specialist resources, administration and systems required. You would have to provide these yourself – at considerable cost – if you ran a dedicated captive.

Existing expertise

Manning & Rorsch already has the necessary skills to assume the complex management responsibilities associated with running a captive.

Lower capitalisation

A Manning & Rorsch cell requires a minimum solvency of only 25% of premium income. A wholly owned captive, on the other hand, requires a capitalisation of at least £50m (for a minimum asset base of £30m).

No need to consolidate

A cell owner doesn't have to consolidate his investment in Manning & Rorsch because the shareholding is so small. A dedicated captive, however, is much larger and would therefore have to consolidate.

5. Notice in the parking lot of a shopping mall

IMPORTANT

IN THE INTERESTS OF SECURITY, PLEASE REMOVE ALL VISIBLE ITEMS OF VALUE (ESPECIALLY REMOVABLE RADIOS AND CELLPHONES) FROM YOUR VEHICLE WHEN PARKING. DO NOT LEAVE PARKING TICKETS OR ACCESS CARDS IN YOUR VEHICLE. ENSURE THAT YOUR VEHICLE IS LOCKED.

What's wrong?

1. Capital letters (difficult to read).
2. Meaningless headline.
3. Legalistic writing style.

Is your car safe?

Remove visible items of value, especially detachable radios and cell-phones

Keep your parking ticket/access card

Lock your car

6. An article for an investment newsletter

Gold

Although the price of gold has followed a downward trend over the past 16 years, it was the announcement from the Australian central bank of sales of 167 tonnes of gold in July of 1997 that initiated the plunge in the gold price to levels last experienced 13 years ago. With this sale, Australia replaced two-thirds of its reserves with US$ denominated 'paper'. Although central banks have been selling gold from their reserves for years, substantial amounts have rarely, if ever, been sold from the vaults of gold producing countries. Australia, being amongst the top five gold-producing countries in the world, triggered the traditional role of gold as a store of value being questioned.

Not long after this, the collapse of the financial systems of the Asian countries brought economic turmoil to the region, taking the world by complete surprise. With the collapse of these economies, demand for luxury goods, including jewellery, would deteriorate, further spurring a fall in the gold price. Capitalising on the massive negative sentiment surrounding demand for gold, "twenty-three-year-old screen jockeys" embarked on an exercise of large-scale short selling reducing the gold price to below the $280 level. Incidentally, the average real gold price measured over the past 100 years is in the region of $285.

The uncertainty currently surrounding gold's recovery to prices required to return the global gold mining industry to profitability essentially hinges on the decision of the European Union (EU) as to the quantities to be retained by its central bank. Encouragingly, these quantities are believed to be equivalent to those currently contained in the vaults of its members as there has been no reporting of central bank sales by Germany, France or Italy, countries that are major holders of gold. The collapse of the Asian economies might further reinforce gold's function as a store of value with which to back up the new EU currency, at least for the foreseeable period of uncertainty and currency revaluation. Although the perception of gold as a store of value seems to be waning among members of the developed Western economies, the increased purchasing of gold by the Russian central bank, and the record demands recently experienced from the Middle East and India, are just two examples which show that gold is still perceived by the developing economies as a store of value.

Negative sentiment surrounding the gold price will remain in place until there is more certainty as to central bank selling. However, there are certain factors that should not be overlooked when deciding on whether the gold price will recover in the short term. The primary factors include: the physical demand (total fabrication) and supply (mine output) fundamentals remain firmly in place, with annual demand substantially exceeding supply, and, the speedy strengthening of the Asian economies. Having faith in the basic fundamentals of economics, and the Asian culture that brought about the rise of the Asian Tigers, gold's recovery appears to be unquestionable.

What's wrong?

1. Non-descriptive headline which begs the question – what about gold?
2. Dense, wordy layout.
3. The point – that gold is far from dead – is buried in the last paragraph.

Don't write off gold just yet

It's far from dead, despite the Asian crisis

Fundamentals look good

Gold has taken a pounding since the Asian crisis – but that's no reason to write it off. Here's why:

- physical demand (total fabrication) and supply (mine output) fundamentals remain firm
- annual demand substantially exceeds supply
- the Asian economies are speedily strengthening.

Still a store of value

The collapse of the Asian economies might further reinforce gold's function as a store of value with which to back up the new EU currency, at least for the foreseeable period of uncertainty and currency revaluation. Also, increased gold purchases by the Russian central bank and record demand from the Middle East and India prove it is still perceived by developing economies as a store of value.

Recovery depends on central bank holdings

Negative sentiment surrounding the gold price will remain in place until there is more certainty about central bank holdings – particularly those of the EU. Encouragingly, it looks as if the EU bank will hold gold equal to what is currently in its members' vaults as there have been no reported central bank sales by major gold holders such as Germany, France and Italy.

The bad news started in Australia

Admittedly, the metal was looking bad after the Australian central bank suddenly sold 167 tonnes in July 1997 (two-thirds of its reserves) and replaced it with US$ denominated 'paper'. This surprise move sent gold plummeting to price levels last seen 13 years earlier and put a question-mark over its traditional role as a store of value.

Then things got worse when Asia collapsed

In late 1997, the financial systems of the Asian countries collapsed. The resulting fall in demand for luxury goods, including jewellery, produced a further deterioration in the gold price and it sank to below the $280 level. (It's worth noting, though, that the average real gold price measured over the past 100 years is in the region of $285.)

7. A set of graphs from an investment presentation

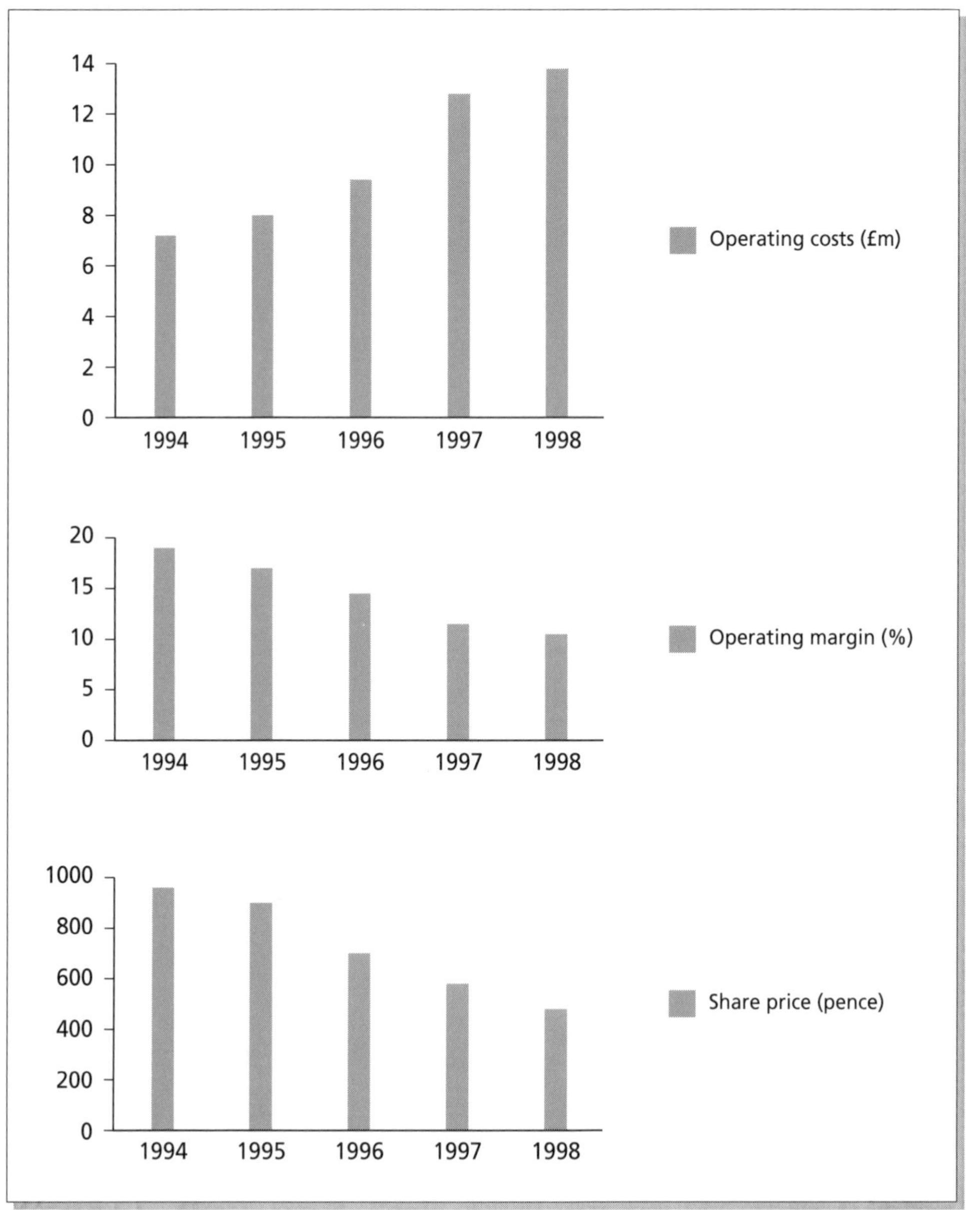

What's wrong?

1 No headlines.
2 No numbers on bars.

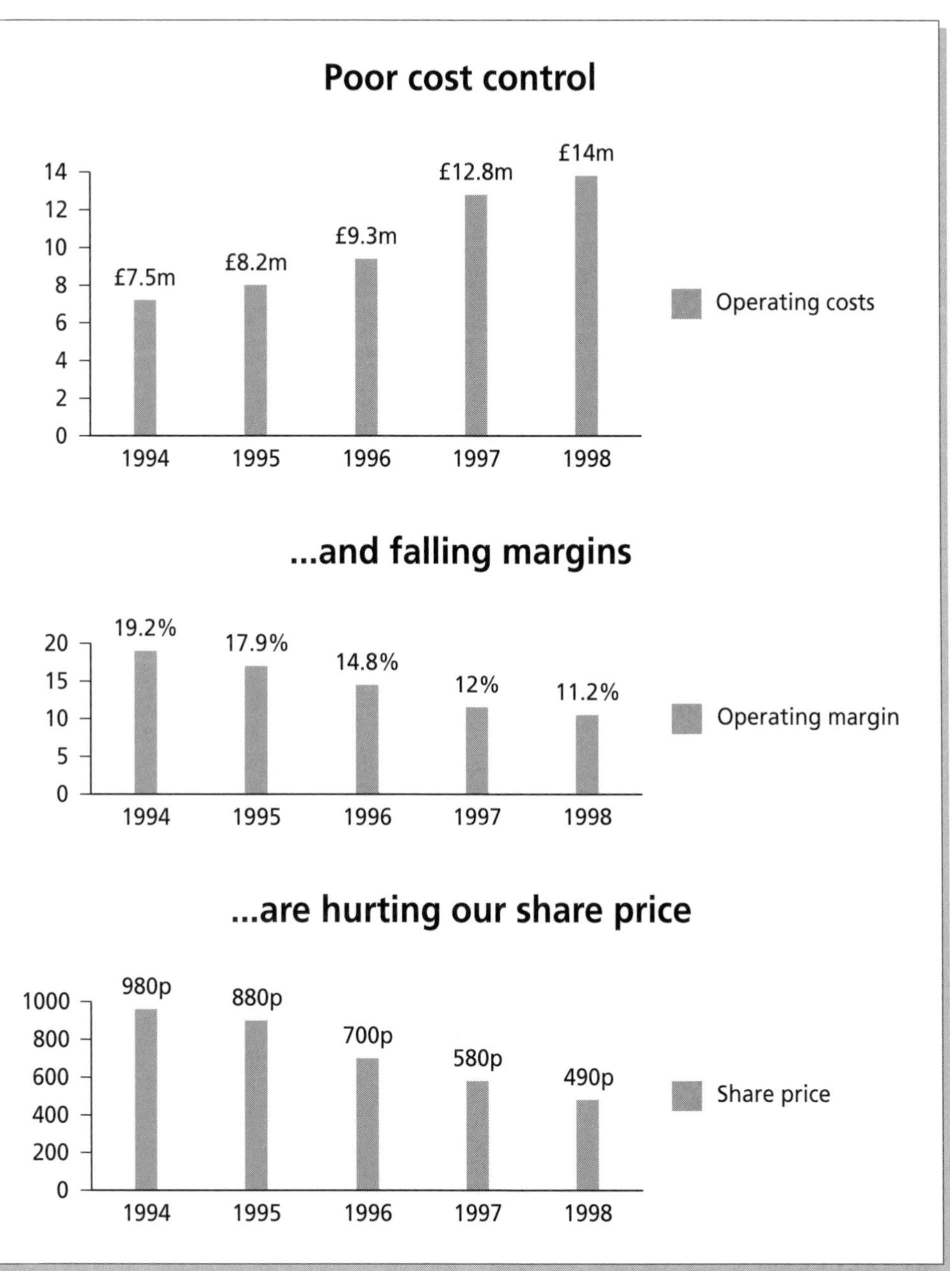
Poor cost control
14
12
10
8
6
4
2
0
£7.5m
£8.2m
£9.3m
£12.8m
£14m
1994
1995
1996
1997
1998
Operating costs
...and falling margins
20
15
10
5
0
19.2%
17.9%
14.8%
12%
11.2%
1994
1995
1996
1997
1998
Operating margin
...are hurting our share price
1000
800
600
400
200
0
980p
880p
700p
580p
490p
1994
1995
1996
1997
1998
Share price

"Many documents fail because they are so ugly that no one will read them, or so confusing that no one can understand them."

KAREN SCHRIVER
Dynamics in Document Design

CHAPTER 2 How to create a user-friendly layout

Clearing up some common myths

Myth: layout doesn't really matter in business writing
Reality: it matters in _all_ kinds of writing
We're used to spending time and money on the visual appeal of a glossy brochure or annual report. Yet doing the same for day-to-day business documents such as letters and reports somehow strikes us as odd. This makes no sense as good design and layout helps us to process *any* kind of written information more easily. It's not enough for your document to be well written; it has to be nicely presented too.

Myth: if it's really important, people will read it
Reality: they may not if it's badly presented
Remember the movie *Field of Dreams*? It gave us the memorable line: *If you build it, they will come*. Most writers of business documents have a similar motto: *If you write it, they will read*. This is a dangerous assumption. Reading is a voluntary process, and we all have a frustration threshold beyond which we simply will not go when faced with poorly presented material.

Myth: we can live with indifferent presentation
Reality: it causes lost sales and reduced productivity
All over the world, a staggering amount of money is lost every year through poorly presented material. Stockbroking analysts struggle through annual reports, commuters can't make sense of railway timetables, taxpayers battle with tax forms and our eyes glaze over when confronted with yet another confusing memo. In a celebrated case reported in 1983 by the *Wall Street Journal*, a US computer company had to withdraw a new home computer from the market after countless users returned it, claiming they couldn't understand the instruction manual.

"For a layout to work, it must get your message across quickly … It must be organized so the reader can move smoothly and easily through the piece."

LORI SIEBERT and LISA BALLARD
Making a Good Layout

Five steps to a user-friendly layout

1 **Structure**
Establish broad meaning: break your text up into clear, logical sections

2 **Differentiation**
Without overdoing it, use different font sizes and styles

3 **Downward flow**
Arrange your text in narrow columns rather than wide ones

4 **White space**
Make your text breathe by having enough white space around it

5 **Focal point**
Create a dominant area of interest that catches the eye

"You have to deliver the information the way people absorb it, a bit at a time, a layer at a time, and in the proper sequence.

If you don't get their attention first, nothing that follows will register. If you tell too much, too soon, you'll overload them and they'll give up. If you confuse them, they'll ignore the message altogether."

PACO UNDERHILL
Why We Buy

Structure

Break your text up into clear, logical sections

This creates a mental structure from which the detail can be understood

Simplest way to do this

Think in terms of headlines, not detail. Picture your reader scanning the document trying to figure it out at a glance. The questions he is most likely to ask reflect the headlines you should be using.

Survey results

"We want more comfort!"

75% of economy-class passengers are prepared to pay for extra legroom

Discomfort the main reason

xxxxxxxxxxxxxxxxxxxxxxxxxxxxxxxxxxxxxxx
xxxxxxxxxxxxxxxxxxxxxxxxxxxxxxxxxxxxxxx
xxxxxxxxxxxxxxxxxxxxxxxxxxxxxxxxxxxxxxx

Similar findings on other airlines

xxxxxxxxxxxxxxxxxxxxxxxxxxxxxxxxxxxxxxx
xxxxxxxxxxxxxxxxxxxxxxxxxxxxxxxxxxxxxxx
xxxxxxxxxxxxxxxxxxxxxxxxxxxxxxxxxxxxxxx

Over 14 000 passengers surveyed

xxxxxxxxxxxxxxxxxxxxxxxxxxxxxxxxxxxxxxx
xxxxxxxxxxxxxxxxxxxxxxxxxxxxxxxxxxxxxxx
xxxxxxxxxxxxxxxxxxxxxxxxxxxxxxxxxxxxxxx

Three reasons to join the Gold Club

1 **A monthly newsletter**
xxx xx xxx xxxxx xxx x xxxx xx xxxxxx xxx xx xxx xxxxx xxx xxx xx xxxxx x xx xxxxx

2 **Daily online investment tips**
xxx xxxxxx xxxxx xxx xxxxx xxx x xxxxx xxx xx xxx xxxx xxx xxxxx xx xxx xxx xxxxx

3 **Big discounts on purchase of mutual funds**
xxx xx xxx xxxxx xxx x xxxx xxx x xxxxx xxxxx xxx xxxx xxx xxx xx xxxxx x xx xxxxx

How to sign up

xxxxxx xx xxxxx x xxx xxxxx xxx xxxxxxx xxxxxxx x xxx xxxxx xxx x xxxxx xxx xxxxxx xx xxxxx x xxx xxxx xxx xxxxxxx xxxxxx xx xxxxx x xxx xxxxx x xxx xxxx xxx xxxxxxx

Get a friend to join and earn 5 000 points

xxxxxx xx xxxxx x xxx xxxxx xxx xxxxxxx xxxxxxx x xxx xxxxx xxx x xxxxx xxx xxxxxx

"Typography is the efficient means to an essentially utilitarian, and only accidentally aesthetic, end, for the enjoyment of patterns is rarely the reader's chief aim."

STANLEY MORISON

Differentiation

Without overdoing it, use different font sizes and styles

This relieves the monotony of the page and makes it easy to pick out key words and phrases

Simplest way to do this

Isolate specific pieces of text (eg, titles, numbers) and highlight them by playing around with:

- boldness of text
- capitals
- italics
- different fonts.

Remember to use capitals sparingly. They're okay in titles and headings, but never in entire sentences.

CALLING ALL MBA HOPEFULS

If you're a senior employee and wish to apply for the Chairman's MBA Scholarship, you must first attend a preliminary interview. The following time slots have been scheduled for the interviews:

Monday	6 March	10h00	Boardroom 1
Tuesday	7 March	14h30	Boardroom 5
Friday	10 March	10h00	Boardroom 1

Calling all MBA hopefuls

If you're a senior employee and wish to apply for the *Chairman's MBA Scholarship*, you must first attend a preliminary interview. The following time slots have been scheduled for the interviews:

MONDAY	6 March	10h00	*Boardroom 1*
TUESDAY	7 March	14h30	*Boardroom 5*
FRIDAY	10 March	10h00	*Boardroom 1*

Technical information

Manufacturing division

Product specs and delivery times
We are the largest supplier of Category A and Category B borehole equipment and guarantee delivery within 24 hours of receiving your order.

Contact: James McGregor
Sales Manager
PO Box 2317
Marshalltown 2107

tel…….. (011) 555-8172
fax……. (011) 555-8174
e-mail…. james@superbore.com

TECHNICAL INFORMATION

Manufacturing division

Product specs and delivery times

We are the largest supplier of *Category A* and *Category B* borehole equipment and guarantee delivery within 24 hours of receiving your order.

Contact JAMES McGREGOR
Sales Manager
PO Box 2317
Marshalltown 2107

tel (011) 555-8172
fax (011) 555-8174
e-mail *james@superbore.com*

"Which typefaces are easiest to read? Those which people are accustomed *to reading ... The more outlandish the typeface, the harder it is to read. The drama belongs in what you say, not in the typeface."*

DAVID OGILVY
Ogilvy on Advertising

Downward flow

Arrange your text in narrow rather than wide columns

The eye finds it easier to process information whose dominant flow is downwards rather than across

Simplest way to do this

Adjust the width of your margins by playing around with your page settings or column formats. (Admittedly, there will be times when space constraints prevent you from achieving a downward flow.)

Customer satisfaction

How our readers rated the service in bars and restaurants
(all scores out of 10)

US	UK	Holland	France	Spain	Austria	Germany
7.9	7.0	6.7	4.4	3.2	2.6	2.5

Customer satisfaction

How our readers rated the service in bars and restaurants *(all scores out of 10)*

US	7.9
UK	7.0
Holland	6.7
France	4.4
Spain	3.2
Austria	2.6
Germany	2.5

TRAVEL TOURS Shop B42 Charleston Mall 5th Avenue Sandton

tel (011) 555-6754 **fax** (011) 555-6700 **e-mail** *info @ traveltours.com*

TRAVEL TOURS
Shop B42
Charleston Mall
5th Avenue
Sandton

tel	(011) 555-6754
fax	(011) 555-6700
e-mail	*info @ traveltours.com*

"The designer should analyse every element he puts into a page. If it helps reading rhythm, he should keep it; if it doesn't, its value is questionable; if it works against comprehension, it should be eliminated."

COLIN WHEILDON
Communicating, or just making pretty shapes

White space

Ensure your text has enough white space around it

This makes your page breathe and helps to eliminate visual clutter

Simplest way to do this

Always hit the ENTER key once or twice before starting a new paragraph or section. You can also adjust your page setup by playing with the width of the margins or different column formats. You'll soon get a feel for what constitutes "the right amount" of white space.

GLOBAL INTERNET OUTLOOK 2002

Published by the United Technology Forum
"The most authoritative and penetrating analysis of where the Internet is heading in the new millennium"
BRAMLEY COMPUTER REVIEW

PRICE: £20 in paperback, £30 in hardback

FREE copies of this authoritative publication are now available in the library on the 13th floor as well as in the Basement Web Café next to the gymnasium. If you'd like permanent copies for your department, please ask your divisional head to purchase them from Di Porter in the library. This is the flagship publication of the world's leading Internet and IT monitoring organisation. It is based on information provided by more than 5 000 private companies and 33 governments. Written in clear, non-technical language and packed with informative graphics and tables, this is a must-read. (**JOHN McCANN** Head of Information Technology)

GLOBAL INTERNET OUTLOOK 2002

Published by the United Technology Forum

"The most authoritative and penetrating analysis of where the Internet is heading in the new millennium"
BRAMLEY COMPUTER REVIEW

PRICE: £20 in paperback, £30 in hardback

FREE copies of this authoritative publication are now available in the library on the 13th floor as well as in the Basement Web Café next to the gymnasium. If you'd like permanent copies for your department, please ask your divisional head to purchase them from Di Porter in the library.

This is the flagship publication of the world's leading Internet and IT monitoring organisation. It is based on information provided by more than 5 000 private companies and 33 governments. Written in clear, non-technical language and packed with informative graphics and tables, this is a must-read.

JOHN McCANN
Head of Information Technology

"In order to attract readers, every layout needs a focal point. Without one, viewers quickly move on."

LORI SIEBERT and LISA BALLARD
Making a Good Layout

Focal point

Create a dominant area of interest that catches the eye

Draw your reader to any piece of text by making it stand out

Simplest way to do this

Play around with:

- boldness of text
- size
- relative position.

This is particularly important in headlines, which are critical visual cues. The best way to draw attention to a headline is with a bold font.

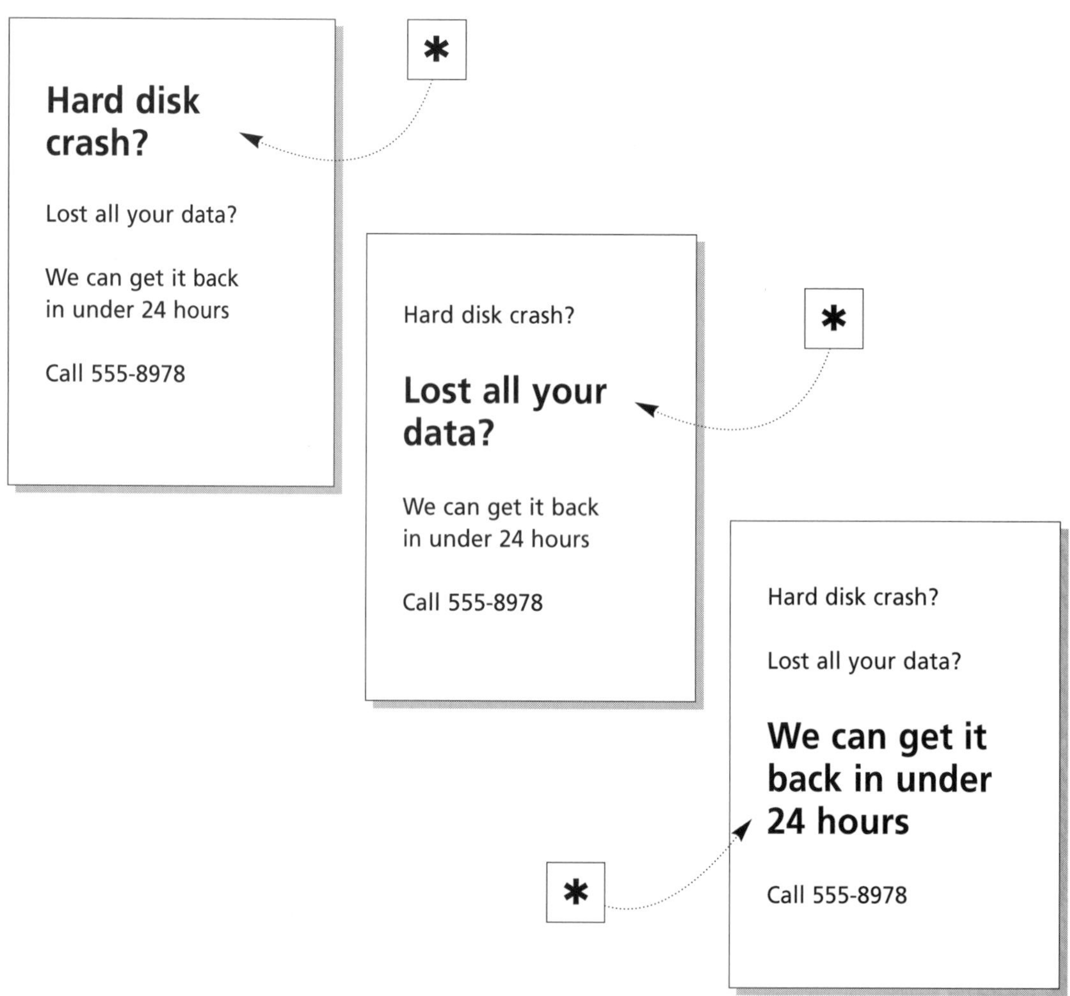
*
Hard disk crash?
Lost all your data?
We can get it back in under 24 hours
Call 555-8978
*
Hard disk crash?
Lost all your data?
We can get it back in under 24 hours
Call 555-8978
*
Hard disk crash?
Lost all your data?
We can get it back in under 24 hours
Call 555-8978

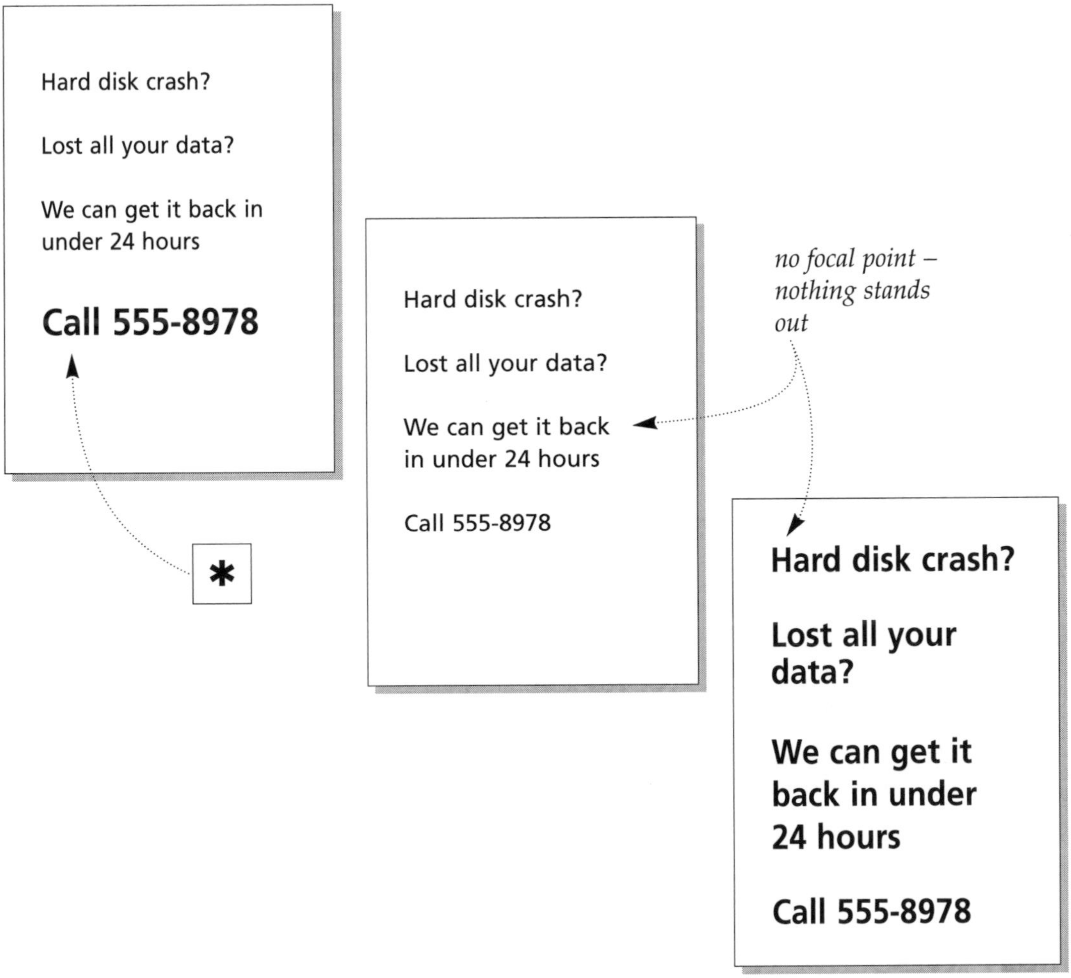

Hard disk crash?

Lost all your data?

We can get it back in under 24 hours

Call 555-8978

✱

Hard disk crash?

Lost all your data?

We can get it back in under 24 hours

Call 555-8978

no focal point – nothing stands out

Hard disk crash?

Lost all your data?

We can get it back in under 24 hours

Call 555-8978

"**Readable.**

1. *Able to be read, legible.* **2.** *Giving pleasure or interest when read; agreeable or attractive in style."*

THE NEW SHORTER OXFORD ENGLISH DICTIONARY

Real-world case studies

Here are six Before/After examples that demonstrate the power of good design and layout:

1. Letter
2. Report
3. Contents page
4. Sign
5. Table
6. Form

It's worth noting that they can all be done in plain MS Word using just two fonts (a serif and a sans serif) and the standard formatting tools available on the screen toolbars.

1. Letter

John Smith
The Really Big Corporation
Sandton Square
West Tower, Suite 52
Sandton 2146

Dear John

List of new employees for UK conference

I'd like the following new people from Marketing and Communications to attend our annual getaway in London in June:

- Terry Laborde age 24 BSc Mechanical Engineering
- Patsy Munroe age 27 BSc Mathematics
- Kelly Wittering age 23 BSc Computer Science
- Bradley Chang age 27 BSc Advanced Physics
- Jane Johnson age 25 BA English, MBA

As you can see, they easily fit our new selection criteria. You may recall that during our last flight together, you were concerned that our UK getaway was turning into a glorified junket for senior executives who were getting a bit long in the tooth.

Well, this lot are anything but: they're young, they're smart and they're hungry. And what's more, they don't yet qualify for business class travel – so we're saving about £10 000 in airfare alone ! No matter which way you slice it, I don't think we could pick a better team. Let me know what you think as soon as you're back from Australia.

Regards

Phillip Terroni
Head of International Marketing

JOHN SMITH
The Really Big Corporation
Sandton Square
West Tower, Suite 52
Sandton 2146

Dear John

List of new employees for UK conference

I'd like the following new people from Marketing and Communications to attend our annual getaway in London in June:

- TERRY LABORDE age 24 *BSc Mechanical Engineering*
- PATSY MUNROE age 27 *BSc Mathematics*
- KELLY WITTERING age 23 *BSc Computer Science*
- BRADLEY CHANG age 27 *BSc Advanced Physics*
- JANE JOHNSON age 25 *BA English, MBA*

As you can see, they easily fit our new selection criteria. You may recall that during our last flight together, you were concerned that our UK getaway was turning into a glorified junket for senior executives who were getting a bit long in the tooth.

Well, this lot are anything but: they're young, they're smart and they're hungry. And what's more, they don't yet qualify for business class travel – so we're saving about £10 000 in airfare alone ! No matter which way you slice it, I don't think we could pick a better team. Let me know what you think as soon as you're back from Australia.

Regards

Phillip Terroni
HEAD OF INTERNATIONAL MARKETING

2. Report

Trade Surplus Beats Expectations Again.

February Surplus Hits R1.9bn On Strong Exports Of Precious Stones, Machinery And Motor Vehicles.

Exports up R2bn to R15.8bn
In February the trade balance once again beat market expectations, showing a surplus of R1.9bn instead of the anticipated R0.8bn. Exports increased by R2bn to R15.8bn, while imports rose by a similar amount to R13.9bn (imports are generally seasonally stronger in February than in most other months).

Third successive surplus

	Dec 99	Jan 00	Feb 00
Exports	R15.6bn	R13.6bn	R15.8bn
Imports	R12.0bn	R11.8bn	R13.9bn
Trade balance	R3.60bn	R1.80bn	R1.90bn

Fig: Trade balance Dec 99 – Feb 00

Precious and semi-precious stones up R0,9bn
Exports of precious and semi-precious stones (mostly diamonds) jumped R0.9bn. Judging by the R0.4bn rise in the other unclassified goods category, commodity exports seem to have risen as well. Large monthly increases were also registered in machinery/mechanical appliances (+ R0.5bn) and vehicles (+ R0.3bn). Exports to Europe and America remained strong, while exports to Africa were lower.

Crude oil largest component of imports
Given the high seasonality of imports and the low level of mineral imports in January, the rise in total imports in February came as no surprise. The rise in imports was fairly broadly based, with the mineral component (mostly crude oil) showing the biggest rise, up R2.4bn.

Cumulative figures encouraging
The trade surplus in February brings the cumulative trade surplus for the first two months of 2000 to R3.6bn. This compares favourably with the R2.3 bn surplus over the same period last year. Measured on the same basis, nominal exports rose 19.6% year-on-year versus a 15.8% year-on-year rise in imports. While imports rose mainly on account of the near doubling in the Rand oil price, other import categories, notably machinery (which is good for economic growth and reflects a pick-up in investment spending), showed healthy growth.

This information is considered to be reliable. However, we do not guarantee its accuracy, nor are we responsible for any decisions based on it.

Trade surplus beats expectations again

February surplus hits R1.9bn on strong exports of precious stones, machinery and motor vehicles.

Exports up R2bn to R15.8bn

In February the trade balance once again beat market expectations, showing a surplus of R1.9bn instead of the anticipated R0.8bn. Exports increased by R2bn to R15.8bn, while imports rose by a similar amount to R13.9bn (imports are generally seasonally stronger in February than in most other months).

Third successive surplus

	Dec 99	Jan 00	Feb 00
Exports	R15.6bn	R13.6bn	R15.8bn
Imports	R12.0bn	R11.8bn	R13.9bn
Trade balance	R3.60bn	R1.80bn	**R1.90bn**

Fig: *Trade balance Dec 99 – Feb 00*

Precious and semi-precious stones up R0,9bn

Exports of *precious* and *semi-precious stones* (mostly diamonds) jumped R0.9bn. Judging by the R0.4bn rise in the *other unclassified goods* category, other commodity exports seem to have risen as well. Large monthly increases were also registered in *machinery/mechanical appliances* (+ R0.5bn) and *vehicles* (+ R0.3bn). Exports to Europe and America remained strong, while exports to Africa were lower.

Crude oil largest component of imports

Given the high seasonality of imports and the low level of *mineral* imports in January, the rise in total imports in February came as no surprise. The rise in imports was fairly broadly based, with the *mineral* component (mostly crude oil) showing the biggest rise, up R2.4bn.

Cumulative figures encouraging

The trade surplus in February brings the cumulative trade surplus for the first two months of 2000 to R3.6bn. This compares favourably with the R2.3 bn surplus over the same period last year. Measured on the same basis, nominal exports rose 19.6% year-on-year versus a 15.8% year-on-year rise in imports. While imports rose mainly on account of the near doubling in the Rand oil price, other import categories, notably *machinery* (which is good for economic growth and reflects a pick-up in investment spending), showed healthy growth.

This information is considered to be reliable. However, we do not guarantee its accuracy, nor are we responsible for any decisions based on it.

3. Contents page

Contents

CONTENTS

4. Sign

NO ENTRY WITHOUT AN ACCESS CARD

THIS IS A RESTRICTED FLOOR. IF YOU DO NOT HAVE AN ACCESS CARD, WE <u>CANNOT</u> LET YOU IN – EVEN IF YOU ARE ACCOMPANIED BY AN EMPLOYEE. ACCESS CARDS ARE AVAILABLE AT SECURITY POINT ALPHA ON THE 2ND FLOOR. THANK YOU.

REFERENCE LOCATION: 17F-3A

No entry without an access card

This is a restricted floor.

If you do not have an access card, we *cannot* let you in – even if you are accompanied by an employee.

Access cards are available at Security Point Alpha on the 2nd floor.

Thank you

Reference location: 17F-3A

5. Table

	OWNERSHIP % STAKE	SALES $ MILLION	PROFIT $ MILLION
ASIA/PACIFIC			
AUSTRALIA	17%	3.78	1.22
INDONESIA	22%	2.34	0.98
MALAYSIA	10%	2.79	0.75
THAILAND	22%	5.77	3.26
SINGAPORE	20%	8.87	3.55
EUROPE			
FRANCE	17%	9.87	2.33
GERMANY	10%	8.34	3.89
HOLLAND	22%	4.44	1.98
BRITAIN	25%	9.99	5.34
NORTH AMERICA			
CANADA	33%	17.2	11.4
USA	44%	33.8	15.8
SOUTH AMERICA			
ARGENTINA	11%	8.33	4.12
CHILE	11%	6.45	0.98
VENEZUELA	13%	3.33	0.44
AFRICA			
BOTSWANA	44%	0.98	0.44
SOUTH AFRICA	17%	13.2	7.40
ZAMBIA	16%	1.34	0.88

Fig – Table of sales and profits from overseas investments broken down by region. (Excludes acquisitions made after 21 November.)

REGIONAL BREAKDOWN

Sales and profits from overseas investments

(excludes acquisitions made after 21 November)

	Ownership % stake	Sales $ million	Profit $ million
Asia / Pacific			
Australia	17%	3.78	1.22
Indonesia	22%	2.34	0.98
Malaysia	10%	2.79	0.75
Thailand	22%	5.77	3.26
Singapore	20%	8.87	3.55
Europe			
France	17%	9.87	2.33
Germany	10%	8.34	3.89
Holland	22%	4.44	1.98
Britain	25%	9.99	5.34
North America			
Canada	33%	17.2	11.4
USA	44%	33.8	15.8
South America			
Argentina	11%	8.33	4.12
Chile	11%	6.45	0.98
Venezuela	13%	3.33	0.44
Africa			
Botswana	44%	0.98	0.44
South Africa	17%	13.2	7.40
Zambia	16%	1.34	0.88

REGIONAL BREAKDOWN

Sales and profits from overseas investments

(excludes acquisitions made after 21 November)

	Ownership % stake	Sales $ million	Profit $ million
Asia / Pacific			
Australia	17%	3.78	1.22
Indonesia	22%	2.34	0.98
Malaysia	10%	2.79	0.75
Thailand	22%	5.77	3.26
Singapore	20%	8.87	3.55
Europe			
France	17%	9.87	2.33
Germany	10%	8.34	3.89
Holland	22%	4.44	1.98
Britain	25%	9.99	5.34
North America			
Canada	33%	17.2	11.4
USA	44%	33.8	15.8
South America			
Argentina	11%	8.33	4.12
Chile	11%	6.45	0.98
Venezuela	13%	3.33	0.44
Africa			
Botswana	44%	0.98	0.44
South Africa	17%	13.2	7.40
Zambia	16%	1.34	0.88

REGIONAL BREAKDOWN

Sales and profits from overseas investments

(excludes acquisitions made after 21 November)

	Ownership % stake	Sales $ million	Profit $ million
Asia / Pacific			
Australia	17%	3.78	1.22
Indonesia	22%	2.34	0.98
Malaysia	10%	2.79	0.75
Thailand	22%	5.77	3.26
Singapore	20%	8.87	3.55
Europe			
France	17%	9.87	2.33
Germany	10%	8.34	3.89
Holland	22%	4.44	1.98
Britain	25%	9.99	5.34
North America			
Canada	33%	17.2	11.4
USA	44%	33.8	15.8
South America			
Argentina	11%	8.33	4.12
Chile	11%	6.45	0.98
Venezuela	13%	3.33	0.44
Africa			
Botswana	44%	0.98	0.44
South Africa	17%	13.2	7.40
Zambia	16%	1.34	0.88

6. Form

RESERVED PARKING

For Broad Street office staff only. To apply for a parking card, complete the form below.

1.	**Your personal details**
1.1	First name
1.2	Surname
1.3	Department
1.4	Telephone
1.5	E-mail

2.	**Your car details**
2.1	Make of car
2.2	Model and year
2.3	Colour

3.	**Your vacation particulars**
3.1	No of vacation days / year (days)
3.2	Length of typical vacation (days)
3.3	Note: These answers will help us plan sharing of parking bays during peak periods

Fax back the completed form to (0134) 344-5534. It will take about a week to process your application. Your card will be delivered to your office.

Reserved parking

(For Broad Street office staff only)

To apply for a parking card, complete the form below

Your personal details

First name
___
Surname
___
Department
___
Telephone
___
E-mail
___

Your car details

Make of car
___
Model and year
___
Colour
___

Your vacation particulars

Number of vacation days / year (days)
___
Length of typical vacation (days)
___

NOTE: *These answers will help us plan sharing of parking bays during peak periods*

Fax back the completed form to: (0134) 334-5534

It will take about a week to process your application.
Your card will be delivered to your office.

"[It has been] estimated that up to 30% of letters and memos in industry and government do nothing more than seek clarification of earlier correspondence or respond to that request for clarification ... Bad writing is bad for business."

MARYANN PIOTROWSKI
Better Business Writing

PART 2 Applying the basics

"All letters, methinks, should be free and easy as one's discourse, not studied as an oration, nor made up of hard words ..."

DOROTHY OSBORNE, 1627–95
The Oxford Concise Dictionary of Quotations

CHAPTER 3 Letters, faxes and e-mail

Clearing up some common myths

Myth: always start with "thank you" when replying
Reality: start with whatever is effective

An effective letter gets to the point in the first sentence, whether it's good news (*Congratulations – you've been accepted at New York University*) or bad news (*We're sorry to hear about your accident, but unfortunately we are unable to give you the reserved parking spot you requested*). Why waste the most precious line of your letter with a thank you which, very often, is not even sincere? Starting with thank you makes sense only if the purpose of the letter is to genuinely thank your reader for something.

Myth: you must use formal language
Reality: plain, everyday language is better

An effective letter makes its point quickly and persuasively. Formal, stilted language just gets in the way of this. Why say *Receipt is hereby acknowledged of your income tax return in respect of the current year* when you can say *We have received your tax return for the current year*? Letters steeped in the humility, deference and formality of the 1950s are totally out of place in today's more relaxed, cooperative work environment. More importantly, they are often ineffective. Like music, clothing and other forms of expression, letter writing reflects the times.

Myth: letters are different from e-mail, faxes or memos
Reality: they're all the same thing – correspondence

What is a fax but a letter that gets to you in less than a minute? What is an e-mail but a letter that gets to you in a split second? And is a memo really that different from a letter? It's all just correspondence – the only real difference is the technology. Why lock yourself into one style for e-mail and faxes, and another for memos and letters? Say whatever you have to say as effectively as possible, and send it off using the appropriate technology. The medium is not the message.

"The first line you write must have absolute bite. You must get your reader in quickly, or else your letter will go straight into the waste paper basket."

ALAN PEASE and PAUL DUNN
Write Language

How to ensure that your letter gets read

Keep it to one page

In today's hectic business environment, your reader is almost certainly snowed under with paper, or drowning in e-mail. The biggest favour you can do him is to keep your correspondence to a single page; he has neither the time nor the inclination to wade through anything longer.

Don't write it like a novel

Don't write your letter like a novel with a teaser introduction, a slow build-up and the point at the end. Many readers find this irritating and a waste of their time. Write your letter like a newspaper article – make your point in the very first paragraph.

Don't state the obvious

When you are replying to someone's letter, you've obviously received it. So try not to open with the sentence *We have received your letter dated....* Similarly, if your reader has to query your letter, he'll contact you. There's no need to always sign off with *If you have any queries, don't hesitate to contact me.*

Get personal – use real language

Your request has been declined. By whom – the janitor? *Acme Incorporated would like to express its apologies.* Is that all the assets of Acme that are apologising, or just the bricks and mortar? You're a real, flesh-and-blood person, not a robot; use real language (I, we, etc) that your reader can relate to.

Avoid tired, stock phrases

Tired, overused expressions such as *at your earliest convenience, as a valued customer* and *I trust the above is in order* suggest that very little original thought has gone into your letter. Your reader has seen it all before.

"We write tight because, like virtue, doing so is its own reward ... But more important, tight writing makes its point – and in its concision and swiftness, delivers information and impact."

WILLIAM BROHAUGH
Write Tight

Five steps to short, persuasive letters

Start with a headline
It tells your reader instantly what your correspondence is about.

Make your point in the first paragraph
Don't bury it towards the end; your reader may not get that far.

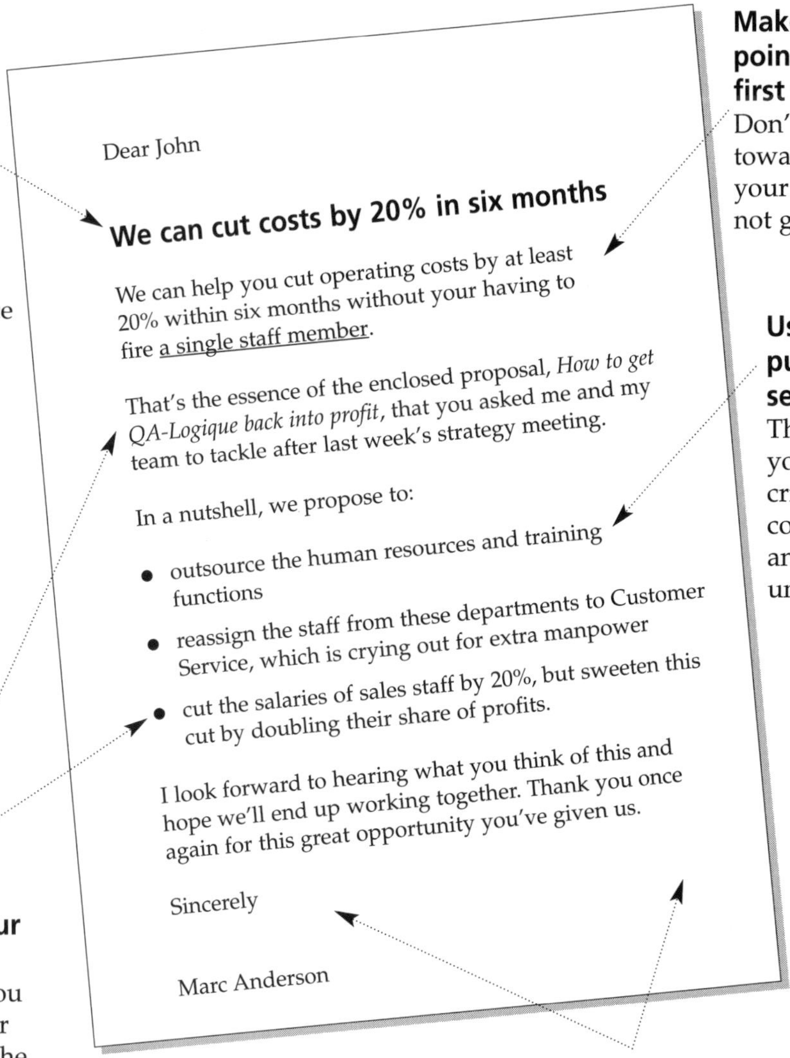

Dear John

We can cut costs by 20% in six months

We can help you cut operating costs by at least 20% within six months without your having to fire a single staff member.

That's the essence of the enclosed proposal, *How to get QA-Logique back into profit*, that you asked me and my team to tackle after last week's strategy meeting.

In a nutshell, we propose to:

- outsource the human resources and training functions
- reassign the staff from these departments to Customer Service, which is crying out for extra manpower
- cut the salaries of sales staff by 20%, but sweeten this cut by doubling their share of profits.

I look forward to hearing what you think of this and hope we'll end up working together. Thank you once again for this great opportunity you've given us.

Sincerely

Marc Anderson

Use short, punchy sentences
They make your letter crisp, conversational and easy to understand.

The body supports your point
That's how you complete your argument in the most persuasive manner.

Use a clean, airy layout
It makes your text stand out and allows you to highlight key points.

"The most important piece of any writing is its focus … When I was working for a newspaper, my editor often urged me to write the headline first. Doing so put the main idea into eight or 10 words and focused my writing on the task at hand."

PHILIP R. THEIBERT
Business Writing for Busy People

Start with a headline

It tells your reader instantly what the letter is about

A headline is just plain common sense

Can you imagine reading a newspaper article that didn't have a headline? Or a book that didn't have a title? Or an e-mail with nothing written in the subject field? Let's face it: nobody likes to read a document without having some idea of what it is going to say. A letter is no exception.

It's hard to find fault with a headline

You may object to a headline on the grounds that it makes your point too soon. You may argue that it's better to build your case and state your conclusion quietly at the end. However, your conclusion is the same in both cases, so you may as well put it at the front – if only to save time.

Make your headline descriptive

A descriptive headline describes what's about to follow. In a letter responding to a client's enquiry about availability of a product, *Stock available for your order* is a descriptive headline; *Stock* isn't. In a memo changing the venue of your AGM (Annual General Meeting), *AGM to be held in Paris* is a descriptive headline; *AGM venue* isn't.

Re is not a meaningful headline

Re, from the Latin word res (which means *in the matter of*) is often used to indicate the subject of a letter. By all means use it, but remember that your reader invariably knows the subject of the letter already; what he's really after is the content. For example, *Re: your application for a scholarship* begs the question – has the application been accepted or rejected?

"A lot of memos would be more persuasive if their first and last paragraphs were switched ... If you have a new way to cut costs, announce it in sentence one. If you're asking for a new computer, say so at the start."

MARK McCORMACK
McCormack on Communicating

Make your point in the first paragraph

Don't bury it towards the end – your reader may never get there

The point is what matters

Does your letter ask for a transfer because you're unhappy in your present position? Then the point of the letter is the transfer, not the fact that you're unhappy. Does your memo instruct staff to clear their desks at night because exposed paper is a fire hazard? Then the point of your memo is to get staff to clear their desks, not to engage in a technical explanation about fire. Start with what you want to achieve (eg, a transfer, clear desks) – then follow up with the supporting argument.

Point upfront makes for productive reading

Imagine a letter that starts *I'd like to request a transfer to our Glasgow office because I really don't get on with the people here.* Because the point of the letter is clear, your reader can now focus on the reasons for this drastic decision. He could even save time by referring the letter to someone else because he doesn't deal with transfers. This is productive reading. Now, imagine that your letter starts with five rambling paragraphs about how unhappy you are: your reader will spend his time wondering where all of this is leading.

Broad meaning first, detail afterwards

Making your point upfront makes neurological sense because that's how the brain functions. When we process information, we look for broad meaning first (ie, the point), then the detail. That's why newspaper articles, with their bold headlines and powerful opening paragraphs, are so easy to read.

" ... I seriously question any memo over one page. If an idea can't be expressed on one page, it (or its author) is probably flawed."

JACK TROUT with STEVE RIVKIN
The Power of Simplicity

The body must support your point

That's how you complete your argument in the most persuasive manner

You've made your point – now elaborate

If the point of your letter is to ask for a transfer, then the body of the letter should talk about the reasons. If the point of your memo is to get staff to clear their desks at night, then the body of the memo should explain why messy desks are a fire hazard. If the point of your e-mail is to get someone to buy advertising in your magazine, then the body should explain why this makes financial sense.

Use a top-down approach

As you develop the body of your letter, use a top-down or inverted pyramid approach. This is just a fancy way of saying that you should put the most important bits first, and the least important bits last. There's no hard and fast rule for this; it's really about stringing your argument together logically. For example, if your memo opens by stating that your AGM has had to be moved to another venue, then the body will probably cover:

- why? (the travel agent double-booked the original venue)
- how? (a computer error)
- when? (we found out last week)
- consequences? (we depart one hour earlier)

Wrap up with something meaningful

If it's a particularly tricky apology letter, you may want to sign off with: *You have my personal assurance this won't happen again.* If it's a no-nonsense complaint letter, you may end with: *We really would like to continue this relationship, but unless your service improves, we may have to take our business elsewhere.*

"The notion of what makes good writing is changing… I remember asking [a top executive] what he felt about his people using first-person pronouns.

Me: Some of your people feel they shouldn't use the first person – I, me, we – in their writing because they'd seem to be giving their opinions. What do you think?

Him: I hire people for their opinions! Personal pronouns are an excellent way for them to express their opinions – to me and to anyone else."

EDWARD P. BAILEY
Plain English at Work

Use short, punchy sentences

They make your letter crisp, conversational and easy to understand

It's not about impressing your reader

Remember that the purpose of your letter is not to impress your reader with your grasp of English; it's to get a point across quickly and effectively. Your reader is not going to stop in mid-sentence, mouth wide open with admiration, and say: Wow! Will you look at that choice of words! That is so Hemingway! All he wants is the point of the letter so that he can act on it – then file it or throw it away. So use plain, punchy language.

Write the way you speak

When we talk, we use normal English; however, put us in front of a sheet of paper or a computer screen, and some dark part of our psyche takes over. *Do this now* becomes *Implement this with immediate effect*. *This is the method we'll try to use* becomes *This is the methodology we shall endeavour to utilise*. Few people talk like this in real life, so we have difficulty relating to it. That's why more and more books on business writing encourage us to write the way we speak.

Make your sentences light and conversational

To make your sentences come alive, don't be scared to:

- use contractions (*I'll* for *I will*, *we've* for *we have*)
- start a sentence with *and* or *but*
- use questions for emphasis (*Is this effective? You bet!*).

Don't overstep the mark, though: there's a fine line between conversational English and bad English. Remember your audience. *Yo, we ain't gonna pay no insurance claim, man!* is not what this section is about.

"What is layout, anyway? On its simplest level, it is whatever goes into the 'look' of the page: something that appears open and inviting probably has good layout; something that appears cluttered and uninviting probably has bad layout."

EDWARD P. BAILEY
Plain English at Work

Use a clean, airy layout

It makes your text stand out and allows you to highlight key points

A dense letter turns your reader off

Dense, unreadable type stretching from margin to margin is a turn-off. A clean, airy page – or screen – on the other hand makes sentences more visible. We take the path of least resistance when we process written information: it's human nature. That's why we are favourably disposed towards text that is clear and easy to decipher.

It's okay to use one-sentence paragraphs

Don't be scared to use a single sentence in a paragraph; it's a great way of making your text stand out.

There is no "right" number of paragraphs

The three-paragraph letter probably represents the ultimate in visual balance, which is why most people favour it. However, there's nothing magical about the number three. If seven paragraphs give you a cleaner layout, use seven. If two get your point across, then use two. Go with what is effective.

And if you really can't get it all on one page?

Life's not perfect – there will be times when one page just isn't enough, especially when you've got to list facts, figures or complex arguments. A good solution here is to "unbundle" your letter into two parts: a one-page cover letter which summarises your main argument, and a second page with the detail (see the example on page 121). Many sales letters, with their acres of mind-numbing text explaining the virtues of a product, would be better off using this approach.

"'Appollonius to Zeno, greeting. You did right to send the chick-peas to Memphis. Farewell.'

This letter, sent by an Egyptian civil servant thousands of years ago, contains all the essentials of a good, modern business letter. Its very informality has a contemporary ring to it. It is simple, clear and concise – yet courtesy has not been sacrificed."

L.A. WOOLCOTT AND W.R. UNWIN
Mastering Business Communication

Real-world case studies

The following case studies represent a fairly typical cross-section of the kind of correspondence we encounter in business.

You'll note as you go through the letters that they could just as easily have been e-mail or faxes. So try not to get too hung up about the exact nature of each piece of correspondence; just focus on the clarity of the content.

1. Letter declining a sales order
2. E-mail requesting permission to distribute articles
3. Fax requesting final payment
4. E-mail of thanks
5. Letter suggesting improvements to a proposal
6. Memo to healthclub members
7. Letter from trustees of a retirement fund
8. E-mail to divisional heads requesting information
9. Letter to shareholders about a takeover bid

1. A letter declining a customer's request for a special order

Dear Mrs Bolton

We have carefully considered your letter of 12 October in which you ask us to ship you 5 000 additional Comfy Fit shirts in time for Christmas.

Insofar as our firms have done business with each other for many years, we should like to accede to your request and grant you the increased volumes. However, the Comfy Fit's collar requires a time-consuming softening process. To meet your order would mean reducing the length of this critical phase of the production process. It would seriously compromise quality – and this is something we are not prepared to do. We therefore, regretfully, have to decline your request. However, we would like to suggest an additional 2 000 Eezy Fit trousers instead. They have a very good sales track record and we have enough of them in stock.

I hope that you will agree to this suggestion and I look forward to continuing to receive regular orders from you.

Sincerely

What's wrong?

1 A painstakingly slow build-up to the "no", which is buried in paragraph two.
2 The second paragraph is too long and should be split up.

Dear Mrs Bolton

Not enough time to meet your order

We regret that we are unable to ship you the 5 000 additional Comfy Fit Shirts you requested for Christmas in your letter of 12 October – there just isn't enough time.

It would mean speeding up our production lines and shortening the time-consuming softening process which gives the shirts their comfortable feel. That would compromise their quality.

However, we are able to ship you an additional 2 000 Eezy Fit trousers instead. They sell very well and stock is available.

Turning down your request was a tough decision, especially as our two companies have been doing business for many years. However, we just cannot compromise on quality – I trust that you understand that.

We look forward to further regular orders from you. In the meantime, please let us know if you'd like the additional 2 000 Eezy Fit trousers.

Sincerely

2. An e-mail to the editor of a trade magazine

SUBJECT: Articles in *Travel Risk Digest* (UK)

Dear Ms Cairns

I represent Tippot Travel Insurance, Australia's leading travel insurance company. We're a recent subscriber to your magazine *Travel Risk Digest (UK)* and thoroughly enjoy the articles. They are timely, interesting and of relevance to anyone in the travel field.

The purpose of this communication is to ask your permission to make our clients – there are about 50 in all – aware of these stories. We would like to e-mail them a weekly summary of the top ten stories of each issue. Obviously, we would clearly identify the stories as coming from *Travel Risk Digest (UK)* and would include your website address.

I'm not quite sure what copyright or legal problems a request like this might raise; I trust they are not insurmountable. I'd be most grateful if you could come back to me on this at *suzanne@travelinsurance.co.au* and let me know whether you are comfortable with the idea.

What's wrong?

1 Meaningless headline in subject field.
2 The point is buried in paragraph two.

SUBJECT: Can we send your articles to our clients?

Dear Ms Cairns

We at Tippot Travel Insurance here in Australia like *Travel Risk Digest (UK)* so much that we want to share the articles with our clients. Are you comfortable with this?

As Australia's leading travel insurance company (and a recent subscriber to your magazine), we'd like to e-mail a weekly summary of your top ten stories to our limited client base of 50. We'd clearly identify the stories as coming from *Travel Risk Digest (UK)* and would include your website address.

I'd be most grateful if you could come back to me on this at *suzanne@travelinsurance.co.au*.

Thank you.

Sincerely

3. A fax requesting payment for a long-overdue account

Dear Mr Brown

Thank you for the £200 cheque that you delivered to our premises yesterday and the accompanying letter requesting an extension of time in which to pay your outstanding balance.

As your account is now nearly three months overdue, we find your present cheque quite insufficient. It is hardly reasonable to expect us to wait a further month for the balance of £583.40, particularly as we invoiced the goods to you at a specially low price which was mentioned to you when you bought them.

Your purchase contract specifically bound you to paying the sum within 30 days. We have had to reschedule your payments twice and even elected to waive the interest due on the first month's outstanding balance.

We sympathise with your difficulties, but hardly need remind you that it is in our customers' long-term interests to pay their accounts promptly so as to qualify for discounts and at the same time build a reputation for financial reliability.

In the circumstances, we hope that in your own interests, you will make the necessary arrangements to clear your account without further delay. If this is not done by midday on Friday 12 June, we shall instruct our lawyers to seek legal redress. We look forward to receiving your cheque for the balance by that date.

Yours sincerely

What's wrong?

The point of the fax – that the deadline is on Friday – is buried in the last paragraph. Mr Brown may not even read that far, given his attitude so far towards settling his account.

Dear Mr Brown

Your account must be settled by Friday

You still owe us £583.40 for goods purchased over three months ago. If this sum is not settled in full by 12h00 on Friday 12 June, we will instruct our lawyers to obtain the money from you.

Thank you for the £200 cheque that you delivered to us yesterday – but it was clearly not enough to cover the outstanding balance on your account.

We do sympathise with your financial predicament. Indeed, we have gone out of our way to accommodate you, rescheduling payments twice and even waiving interest on the first month's outstanding balance.

However, the goods you ordered were on sale at an unusually low price – and the contract you signed specifically committed you to prompt payment. We therefore cannot grant you any further extensions.

We expect your account to be settled in full by Friday.

Sincerely

4. An e-mail congratulating a company on a really successful function

SUBJECT: Getaway

Dear Bill

You are to be congratulated for hosting such a superb year-end getaway. Not only did we have a good time; we also made some valuable contacts, one of whom I shall be visiting in Toronto next month.

The cuisine was excellent and the after-dinner stand-up comedian really funny. I also relished the opportunity of playing on the newly completed golf course, a sentiment shared by many of the delegates.

I think it fair to say that your company has developed something of a reputation when it comes to hosting an event such as this. Thank you for inviting me and my team; we used the time well and found it very productive.

Sincerely

What's wrong?

1 Meaningless headline in subject field.

2 Given the subject matter, a lighter, more colloquial tone might be more appropriate.

SUBJECT: Great year-end getaway – thanks!

Dear Bill

Congratulations on a superb year-end getaway!

The food was excellent, the comedian had us in stitches and that golf course was a dream. Your company knows how to throw a good party!

We made some really good contacts – in fact, I'm flying out to Toronto next month to meet one of them.

Thank you so much for inviting me and my team; it really was time well spent.

Sincerely

5. A letter suggesting improvements to a draft tax proposal

Dear Mrs Thornton

Your letter dated 7 August 2002, requesting feedback on your draft tax proposal, refers.

I must thank you for the opportunity you have given me to comment on your draft tax proposal and offer suggestions for improvement. To be offered this opportunity is indeed an honour – all the more so given the scope of the proposal and the impact it will have if its recommendations are followed.

I found the proposal to be fairly forthright in its core arguments, and I particularly liked the Before/After examples. I do hope that the document will achieve its intended aim.

If I might, however, make the following suggestion: headlines are acknowledged to be effective in highlighting information that readers might otherwise miss. The judicious use of headlines would therefore almost certainly make the document easier to read. This is all the more so as the key clauses discussed do not apply to all your readers; headlines would enable them to skip such clauses.

Please find attached to this letter pages 9 and 10 of your original proposal, which I have rewritten to show you what I mean. I have also taken the liberty of enclosing an excerpt from a similar document I wrote recently.

I trust my suggestions are helpful. Please do not hesitate to contact me for any further assistance.

Sincerely

What's wrong?

After much outpouring of gratitude, the writer, on bended knee, humbly makes his suggestions way down in paragraph four. He should have been more direct – after all, his opinion was asked for.

REF: *Your letter of 7 August 2002 inviting me to comment on your tax proposal*

Dear Mrs Thornton

More headlines needed in proposal

I'd like to suggest that you adopt a more headline-driven approach to your draft tax proposal so that readers can catch key points at a glance.

Right now, they are forced to read every clause and sentence to get at the meaning – even though, invariably, only certain clauses apply to them as taxpayers.

I've rewritten pages 9 and 10 to show you what I mean. I've also included an excerpt from a similar proposal I wrote recently.

Apart from that, I think it's a great document. I hope it gets the response you and your supporters are looking for.

Thank you for asking me to comment on the proposal. I am grateful for the opportunity to play a part in this important venture.

Sincerely

6. A memo to members of a health club

ATTENTION

DEAR MEMBER
Lockers have been provided for all the members to use on a daily basis. Unfortunately, due to health and security reasons, members are not allowed to leave goods in a locker overnight.

HEALTH REASON
Should a member place dirty or wet clothing in a locker and not come to the club for a week, this would not only be unhealthy but cause a smell in the area.

SECURITY REASON
At night when the club is being cleared, if lockers are still locked it is impossible for the Manager on Duty to tell what the contents of the lockers are. This poses a major security risk.

We would be most grateful if you would assist us in alleviating this problem by not leaving any lockers locked overnight and leaving the lockers unlocked.

Should any lockers be left locked overnight, they will be secured with a cable and after a week, the contents will be removed from the lockers.

Thank you for your understanding.

Yours sincerely

What's wrong?

Too detailed and elaborate – takes forever to get to the point.

Lockers must stay open overnight

Our lockers are for daily use only. Please remove the contents before you leave the Club today.

Lockers left locked overnight are a security risk (we don't know what's in them) and a health risk (wet, dirty clothes smell !).

Any locker found locked will be secured with a cable and the contents returned to you after a week.

Thank you for your understanding

7. A letter from the trustees of a retirement fund to members

Dear Member

MARKET-LINKED VERSUS GUARANTEED PORTFOLIOS

This letter is to make you aware of the possible implications of having your monies invested either totally or partially in the Megatell Pension and Provident Funds' market-linked portfolios when you near retirement.

The possibility of a large capital loss is very real, as many retirement fund members learnt after the October 1998 market crash. Although the Megatell Pension and Provident Funds' market-linked portfolios are well-diversified, the consequences of a major market downturn would almost certainly result in losses for members retiring around the time of a crash.

If you have all or most of your assets invested in the market-linked portfolios, you are exposing yourself to this possibility. Depending on whether you have other retirement investments, sustaining a capital loss may prove disastrous if you are planning to retire early (from age 55) or are approaching normal retirement age (65).

As an example, on a $100 000 investment, a member invested in the market-linked portfolio would have sustained a large capital loss of up to $25 000 had he retired in August or September 1998, when the markets were in a downturn. This was indeed the case for many retirement funds over this period. Had this member been invested in the Guaranteed Portfolio, however, such losses would have been avoided.

If most of your overall retirement savings are invested in the Megatell Funds and you are within five years of retirement, the Trustees recommend that you invest the bulk of your investments in a Guaranteed Portfolio. If you wish to retain an element of exposure to the market, you may elect to transfer 75% of your assets to the Guaranteed Portfolios and the remaining 25% may be invested in the market-linked portfolios.

The Trustees urge that you give this matter considerable thought, even though a lot depends on your individual circumstances and any other investments you may have. Please contact Jane Mintoff on (011) 555-5555 at Leadbetter Actuaries, the company that handles your retirement fund. She will guide you through your options and, if need be, send a representative to see you.

Sincerely

What's wrong?

1 Long and difficult to follow.
2 No clear focus.
3 Action to be taken buried in last two paragraphs.

Dear Member

Are you 5 to 10 years from retirement?

If so, you should consider adjusting your portfolio now to avoid a capital loss on your savings

The problem

If the stock market should be in a downturn when you gain access to your Megatell retirement fund savings, you could lose a substantial portion of that capital – up to 25% or more.

We regard this as a very real possibility given the current ups and downs in the stock market. For example, many members who retired during the October 1998 stock market crash lost up to $25 000 of every $100 000 of their retirement fund savings.

The solution

If all or most of your assets are currently invested in Megatell Pension and Provident Funds' market-linked portfolios, we recommend that you reduce the risk of a market downturn by gradually switching the bulk of your assets – perhaps 75% of them – into a Guaranteed Portfolio.

You can then leave the rest in the market-linked portfolios, which are well-diversified. This will allow you to maintain an element of exposure to the stock market in case it is doing very well when you retire.

What you should do now

Although a lot depends on your individual circumstances and other investments you may have, we strongly suggest you consider these options if you intend to retire in the next five to ten years.

Please contact Jane Mintoff on (011) 555-5555 at Leadbetter Actuaries, the company that handles your retirement fund. She will guide you through your options and, if need be, send a representative to see you.

Sincerely

8. An internal e-mail requesting urgent information from divisional heads

SUBJECT: Christmas presents

It's that time of the year when the Corporate Affairs Department has to prepare Christmas presents for the company's top clients. Top clients are defined as any client, existing or potential, that you regard as being valuable to your division's revenues.

The presents will be prepared by us in the Corporate Affairs Department, and will be accompanied by cover letters signed by the relevant Divisional Heads. So we shall need you to sign these before the presents can be properly packaged.

Please make sure you send us, by e-mail, the names and addresses of your top clients by Friday 25th November. We'll then prepare the cover letters and hand them back to you for signature on Tuesday 29th November at 10h00.

We'll come and collect the signed letters no later than 12h00 that day.

Please let's try and get it right this year and meet all key deadlines. We can't afford a repeat of last year's fiasco when the presents reached clients in January !

Thanks for your cooperation.

Carolynne Thompson
Corporate Affairs Department

What's wrong?

1 Meaningless headline in subject field.
2 Takes too long to get to the action required.
3 Action steps required not clearly presented.

SUBJECT: Tell us who your top 5 clients are – by Friday !

SUMMARY
Calling all divisional heads ! We need the names and addresses of your five most valuable clients by Friday 25th November. Why? So that we can prepare the cover letters which will accompany their Christmas presents.

DETAIL
Please follow the step-by-step instructions below so we can get the presents out on time.

Step 1
E-mail us the names and postal addresses of your top five clients by Friday 25th November. We'll use them to prepare the cover letters.

Step 2
We'll return the completed cover letters to you on Tuesday 29th at 10h00. Please sign them no later than midday and leave them at your reception, where we'll collect them.

We'll put the signed cover letters and presents into carefully prepared Christmas packages and post them to your clients no later than Wednesday 30th November.

With a bit of luck, we won't have a repeat of last year's fiasco when our Christmas presents arrived in January !

Thanks for your cooperation.

Carolynne Thompson
Corporate Affairs Department

9. A letter by one firm to the shareholders of another firm

Dear Lifescheme Shareholder

Condor-Pitman is pleased to extend an offer to acquire 50% of your shareholding in Lifescheme to be settled by the issue of 125 Condor-Pitman shares for every 25 Lifescheme shares. The offer is subject to receiving valid acceptances for more than 40% of the total Lifescheme shares. Furthermore, you may also sell to us all, or a portion, of the remaining 50% of your Lifescheme shares.

Condor-Pitman is a dynamic, highly successful investment company operating in the lower-income market. Since listing in 1993, Condor-Pitman's market capitalisation has grown from $110m to $6.4bn. The company's share price has risen 25% this year and, according to the share recommendations of several leading analysts, this level of growth will run at a minimum of 30% over the next three years.

Condor-Pitman wants to remain highly focused in the rapidly growing entry-level life assurance niche. Most customers in this country, including Condor-Pitman's, are moving up the income ladder. It is important for Condor-Pitman to retain its substantial, valued and upwardly mobile client base as they require more sophisticated life assurance and investment products – which are currently being sold by Lifescheme. Most importantly, we are convinced of our ability to add substantial value to Lifescheme. Why else would we pay a 37% premium? (If our offer is not successful, the Lifescheme share price will probably fall to its previously low levels. This will represent a missed opportunity for you to get some value out of your present Lifescheme shareholding.)

Condor-Pitman, as a strong, committed, directional shareholder, will maximise the value of the Lifescheme brand in many ways. These efforts, which essentially involve us increasing Lifescheme's sales, improving its computer systems and reducing its high costs, have already been explained to you in previous communications.

The above points clearly argue in favour of your accepting our takeover offer – and we sincerely hope you will do so. Please complete the blue form at the back of the Lifescheme circular and post it together with your Lifescheme scrip in the prepaid and self-addressed envelope.

We thank you for your participation and look forward to working for you, our likely new and valued shareholder.

Sincerely

What's wrong?

Far too complex for the average shareholder. The letter should use plain language and focus on why the deal makes financial sense; the detail can be relegated to a separate section.

Dear Lifescheme shareholder,

Three good reasons to accept our offer

Here are three good reasons why we believe you should accept the offer that Condor-Pitman is making for Lifescheme:

1 **You can make a lot of money in Condor-Pitman shares**
In exchange for your Lifescheme shares, you get five times as many Condor-Pitman shares, which are among the most valuable on the stock exchange. They have risen 25% during 1998 and several analysts estimate annual growth of at least 30% over the next three years.

2 **You can boost the value of your existing Lifescheme shares**
The Condor-Pitman takeover is designed to boost Lifescheme's growth prospects. This will boost your remaining Lifescheme shares.

3 **It is an opportunity that may not come around again**
Unless enough Lifescheme shareholders accept the offer, it will lapse – and the Lifescheme share price will fall back to where it was before the offer. This would represent a missed opportunity.

If you're still unsure or need more information on the takeover, please read the enclosed, one-page summary entitled *The takeover in a nutshell.*

If you do wish to accept our offer, simply complete the blue form at the back of the Lifescheme circular and post it with your Lifescheme shares in the pre-paid envelope before 22 November.

Thank you for your time. We look forward to having you as a future Condor-Pitman shareholder.

Sincerely

The takeover in a nutshell

1 Why do we want to buy 50% of Lifescheme?

We're a highly successful investment company operating in the lower-income market. We want to give our rapidly expanding client base access to the sophisticated life assurance products that Lifescheme sells.

2 How will we boost growth at Lifescheme?

We intend to channel huge amounts of our own business through Lifescheme, give Lifescheme access to our sophisticated computer systems and bring down Lifescheme's high cost structure.

3 Why do we think we can do it?

- Lifescheme badly needs new business – our customers can supply it.
- Lifescheme has outdated computer infrastructure – we have one of the most advanced IT set-ups in the country.
- Lifescheme needs to trim its high cost structure – our unit costs are the lowest in the industry.

4 How does the share offer work?

You can sell us all, or part, of the Lifescheme shares that you now hold. For every Lifescheme share you sell us, you'll get 5 Condor-Pitman shares.

Need more detail?

All these points are explained fully in Chapter 3 of your Offer Document.

"One day, a manager tossed a report on my desk. He said: 'I keep getting reports like this from my staff. I have to reread the damn things over and over just to see what they're trying to say … Plus, they're way too long. Why can't they say the same thing in one page, instead of three?'"

PHILIP THEIBERT
Business Writing for Busy People

CHAPTER 4 Reports

Clearing up some common myths

Myth: reports must be complex and erudite
Reality: short and simple is better

A report is a means to an end, not an end in itself. It has no inherent value over and above the message it's trying to get across. The more quickly and easily your reader grasps that message, the better. Reports that get bogged down in length, complexity and literary pretentiousness are a waste of everyone's time. We are all far too busy.

Myth: people always read a report all the way through
Reality: most people prefer to stop at the summary

Do *you* read every report you get from cover to cover? Well, neither does the average reader. We pick up a report not to savour its prose and structure, but to discover what it requires us to *do* – eg, buy more Coca-Cola shares, reorganise the department or buy a new water cooler. You don't need to read the entire report for that any more than you need to read an entire newspaper article to get the gist. While detail is often vital in reports (you may need to refer to it later), a lot of it is unnecessary; in some companies, it is known as CYA ("cover your ass") stuff.

Myth: the right place for the conclusion is at the end
Reality: readers want the conclusion upfront

If you are reading the sports page, where do you want to see the result of yesterday's cup final – upfront in the headline or way down in the last paragraph? Reports are no different: your reader wants the conclusion right away, not on page 11 after thousands of words describing why the report was written and the intricacies of the methodology. This is, without a doubt, the *biggest single gripe* executives have about reports.

"When a reader picks up a report, his or her first concern is with the bottom line, *or …* What's the score? *It's difficult to believe anyone would fail to give the big picture first. Why, then, are so many reports organised* backward?*"*

ROBERT GUNNING and RICHARD A. KALLAN
How to Take the Fog out of Business Writing

Why many reports aren't read

They are too long

A one-page report has a near 100% probability of being read. With a two-pager, this drops to around 75%. Anything longer, and the likelihood that a report will be read falls rapidly to zero. The reason is straightforward: a thick report requires more time, energy and concentration than most readers are willing to give it.

They aren't user-friendly

Take a look at this book (which is really just a long report): it has a meaningful title, a table of contents, clearly identifiable chapters, helpful headlines, a clean layout and an obvious thread running through it. You can breeze through it, stopping wherever you like. Now pick up a report from your in-tray and what do you see? Probably a sea of words with little attempt at user-friendliness.

The point is not apparent right away

Have you ever wondered why most people flip straight to the back of a report? They're looking for the conclusion, which is invariably buried there. A report is not a novel, which is written sequentially so that we may enjoy the gradual unfolding of the plot. A report is functional and seeks to put a point across. State that point right away and use the rest of the report to back it up.

They use complex, convoluted language

Here's a sentence from a report issued by a leading international audit firm: "This study is based on the premise that the leverage by people can only be enhanced if their contribution to business success is viewed in an integrated, systemic and holistic manner from an interventionist perspective." Now imagine page after page of this verbal quicksand.

"Without wasting paper unnecessarily, you should allow generous margins and reasonable space between columns ... It is no tragedy if a section of a report ends halfway down a page, or if the back of a leaflet has to be left blank."

MARTIN CUTTS
The Plain English Guide

Five steps to short, punchy reports

1 **Have a meaningful title**
It tells your reader instantly what the report is about.

2 **Have a table of contents**
This breaks the document down into logical, digestible sections.

3 **Make your point right away**
That's what your reader is after – so don't bury it towards the end.

4 **Use lots of headlines**
They pull you through the report, allowing you to catch key points at a glance.

5 **Use a clean, airy layout**
It makes your text stand out and allows you to highlight key points.

Meaningful title

Goodbye jargon, hello clarity

The SEC now requires prospectuses to be written in plain language[1]

by
JOHN SMITH
Executive Director
Conquest Investment Bank

June 1998

Have a meaningful title

It tells your reader instantly what the report is about

"Report" is not a title

Have you ever seen a book entitled *Book*? Or a newspaper article entitled *Article*? Why, then, does the word *Report* feature so prominently on the cover of reports? Some proposals carry nothing but the word *Proposal*. These empty words merely state the nature of the document, which is pretty apparent anyway; it's the content that we should be highlighting in the title.

A good title is descriptive

If your report is on the fact that new government legislation will increase your company's medical insurance costs by 50%, you could call it *New legislation will push up our medical insurance costs by 50%*. If your report is about how smoking in your company reduces the efficiency of your ventilation system, you could call it *How smoking reduces the efficiency of our ventilation system*. These titles help the reader by describing what the document is about.

Your title should contain a complete thought

Another way of looking at a descriptive title is that it contains a complete thought. So if your report is on the need for people not to work through lunch because it's bad for productivity, you could call it *Why working through lunch is bad for productivity* or simply *Working through lunch is bad for productivity*. The following titles, on the other hand, represent incomplete thoughts: *Working through lunch* (what about it?), or *Productivity in the workplace* (what about it?) or *Lunch and productivity* (what about them?).

Table of contents

Contents

Have a table of contents

It breaks the document down into logical, digestible sections

A table of contents shows logic and organisation

A table of contents divides your report into manageable pieces and makes it easier to digest. It is always easier to process information that is logically arranged, rather than information that is – or appears to be – scattered about randomly.

Make your chapter headings descriptive

The same points about titles apply to chapter headings – ie, they should be descriptive and contain a complete thought. To illustrate this, let's consider possible chapter headings for a report entitled *How smoking reduces the efficiency of our ventilation system.*

Chapter 1	Let's start a no-smoking policy right away
Chapter 2	Smoke pollutes the ventilation ducts
Chapter 3	Smoke affects the air-conditioners
Chapter 4	Smoke raises the system's maintenance costs
Chapter 5	Most staff favour a non-smoking environment

What about reports with a set format?

When your report has to carry specific sections such as *Findings* and *Recommendations*, make them come alive. Example:

Findings	– Our costs are soaring – Staff morale is low
Recommendations	– Cut back on travel perks – Introduce an incentive scheme

What about short reports?

Reports of a page or two seldom need a table of contents. You should be able to organise the content logically through headlines and by making your point upfront.

Point upfront

Executive Summary

The Securities and Exchange Commission (SEC), the US financial markets watchdog, now requires prospectuses to be written in plain language. The benefits: clearer information, fewer queries and lower printing costs.

The SEC unveils new plain language rules

From 1 October 1998, US prospectuses must be written using plain language principles such as short sentences and concrete, everyday language. They should not contain jargon or highly technical business terms. Also, prospectuses should be designed so that they highlight important information for investors. Pictures, charts, graphics and other features may be used.

Both investors and issuers benefit

Both investors and issuing companies benefit from more readable prospectuses. Investors get information they understand, which prompts them to read it and gives them greater confidence in their investments. Companies incur lower printing and distribution costs because plain language prospectuses tend to be shorter than "traditional" prospectuses; companies also save valuable executive time because of fewer queries from investors.

Investors not reading prospectuses

Prospectuses had become so long and complex that few investors were reading them. This deprived them of one of the cornerstones of investor protection – full and fair disclosure. This take-it-or-leave-it attitude is illustrated in a *Washington Post* cartoon that shows a man and a woman discussing investments over a table covered with papers. The man cracks open a prospectus and says: "You know, I met a guy once who actually read one of these."

Rest of world will have to follow

Other aspects of the global plain language revolution have already reached major global financial centres – for example, clearer contracts and insurance policies. It's just a matter of time before plain language prospectuses arrive as well. This will become all the more urgent as privatisation increases the number of ordinary citizens who have to read and understand prospectuses.

Make your point right away

That's what your reader is after – so don't bury it towards the end

Why this is so important

Making your point as early as possible is critical – this cannot be repeated often enough. It allows your reader to:

- understand the report quickly
- act on it quickly – or pass it on to someone else
- read the rest of it productively, instead of wondering where it is heading.

It is one of the more enduring myths in business that people are prepared to plough through a document in order to discover its message.

Long reports need a summary

Any report longer than a few pages needs a summary; this can be as short as one paragraph or as long as one page. Write it on the assumption that the reader won't read any further; that way, you'll focus on what he *needs* to know, rather than what you'd *like him* to know. And don't let anyone tell you that it can't be done on one page; there are books hundreds of pages long that carry very informative summaries on the back cover.

For a short report, your point is the main headline

For a report that is one or two pages long, your point is invariably the main headline – and whatever sub-text you may think is necessary. (At the top of this page, the sub-text is *That's what your reader is after – so don't bury it towards the end.*)

The benefits of plain language prospectuses

Investors get information they understand and companies incur lower costs

Clearer information for investors

Based on its pilot programmes, as well as letters from the public, the SEC says writing prospectuses in plain English should:

1 Allow investors to make better-informed assessments of the risks and rewards of investment opportunities.
2 Reduce the likelihood that investors will make investment mistakes because of incomprehensible disclosure documents.
3 Reduce investors' cost of investing by lowering the time required to read and understand information.
4 Increase consumers' interest in investing by giving them greater confidence in their ability to understand it.
5 Reduce the number of costly legal disputes because investors are more likely to understand disclosure documents.

Lower costs for companies

Clearer prospectuses will lower printing and distribution costs for issuers because plain English tends to reduce document length; it shortened by an average of 11% for pilot participants. Even if printing and distribution costs dropped by only 5%, the SEC found, firms would save approximately $3 160 per filing. Firms will also spend less time answering questions from investors.

Hard at first – then it gets easier

On average, companies will probably need about 60 additional hours to file a plain language prospectus. However, most participants predicted both cost and hours would fall to current levels as firms gain experience with plain English. This is consistent with the findings of the American Society of Corporate Secretaries. The 12 member companies that had prepared at least one plain English document predicted "no material change" in the average annual time necessary to do the prospectus.

Descriptive headlines

Use lots of headlines

They pull you through the report, allowing you to catch key points at a glance

Why headlines matter

Headlines summarise blocks of text on your page, and that makes the text easy to understand. Indeed, a good report is really nothing but a series of embedded summaries: the title summarises the report, the chapters summarise the sections and the headlines summarise the text.

Make your headlines descriptive

As we've seen already, a headline should be descriptive – ie, it should describe what's about to follow. In an economic report, *Monetary policy* is not a descriptive headline; *Monetary policy under attack* is. In an environmental report, *Pollution levels* is not a descriptive headline; *Pollution levels at record lows* is.

Headlines can unscramble complex text

Headlines are particularly useful in unscrambling long or complex paragraphs so that key points stand out. For example:

Before

A concern expressed was whether writing documents in plain English would impose substantial costs on public companies. The response to this was that while there may be additional costs initially, these would, in all probability, be modest. Moreover, they would diminish as firms learn plain English principles.

After

Concern: Will writing documents in plain English not impose substantial costs on public companies?

Reponse: There may be additional costs, but these will probably be modest and should diminish as firms learn plain English principles.

Clean, airy layout

SEC's rules for plain language prospectuses

They call for common-sense design, clear language and fewer technical business terms

Rules for key sections of the prospectus

Writers must use plain English principles in the front and back cover pages, the summary, and the risk factors sections. The language used must comply substantially with six basic principles:

1. Short sentences.
2. Definite, concrete everyday language.
3. Active voice.
4. Tabular presentation or bullet lists for complex material whenever possible.
5. No legal jargon or highly technical business terms.
6. No multiple negatives.

The document should be designed so that it highlights important information for investors. Pictures, charts, graphics and other design features may be used to make it easier to understand.

Rules for the entire prospectus

Writers must use the following techniques for the entire prospectus:

1. Present information in clear, concise sections, paragraphs and sentences.
2. Whenever possible, use short explanatory sentences and bullet lists.
3. Use descriptive headings and sub-headings.
4. Avoid frequent reliance on glossaries or defined terms as the primary means of explaining information. Terms may be defined in a glossary or other section of the document only if their meaning is unclear from the context. A glossary should only be used if it facilitates understanding.
5. Avoid legal and highly technical business terminology.

Use a clean, airy layout

It makes your text stand out and allows you to highlight key points

Go for lots of white space

Lots of white space is important in making a report inviting and user-friendly. It makes headlines and graphics stand out, and enables you to draw attention to key paragraphs. In other words, it makes your document *readable* – and that's what it's all about.

Why white space enhances readability

Chances are you've already read all the quotes in this book. That's no accident. Each quote is sitting there all by itself on a single page – you can't miss it. Clearly, therefore, the more white space around a piece of text, the easier it is to find. Now if you want to hide a sentence, stick it on a page that is chock-a-block with text!

Paper is a means to an end – relax !

One of the hardest lessons for business writers to learn is that paper is hardly ever wasted if it improves the readability of a document. Remember that paper is often the smallest variable in the overall cost of producing a report; what really matters is executive time – for both writer and reader – and whether the report achieves its intended aim. Keep your eye on the prize, not the price.

"Executives don't want diaries. They want writing organised in a deductive *pattern – conclusions first, followed by the evidence that led to those conclusions."*

ROBERT GUNNING and RICHARD A. KALLAN
How to Take the Fog Out of Business Writing

Real-world case studies

The following reports were written using the techniques taught in this book. Interesting observation: not one of them contains the words *Introduction, Findings* or *Conclusion* – yet their central message is immediately apparent from the descriptive headlines.

1 How companies save money using plain language
2 Implications of a court ruling for a company's business
3 Results of a reader poll for a corporate newsletter
4 An assessment of a company's investment potential
5 A view on the global economy
6 An entry for a corporate award

Case study 1

A report explaining how companies save money by using plain language

Short and sweet. No preamble, no fuss. Note how the headlines neatly summarise each point.

How plain language saves you money[2]

You get fewer queries, fewer errors and waste less time

1 Phone enquiries plummet

US computer company Allen-Bradley simplified the user manual for its programmable computers. Phone enquiries through the call centre fell from 50 per day to 2 per month.

2 $375 000 saved in phone calls

General Electric simplified its software manuals. Users of the new manual made about 125 fewer calls per month. This translated into savings of between $22 000 and $375 000 per year.

3 Productivity gains of $400 000

Federal Express simplified all ground-operation manuals. The average search time for information dropped 28% and the chances of readers finding the correct answer rose by 50%. The money saved in the first year was at least $400 000.

4 Error rate down from 55% to 3%

After UK Customs and Excise simplified its lost baggage forms, the error rate on returned forms fell from 55% to 3%. The new form saves £33 000 a year in staff time.

5 Police slash thickness of procedures book[3]

The Greater Manchester Police Department reduced its procedures book from 750 000 words to 60 000 words by rewriting it in plain language. The result: quicker turnaround time on cases.

6 New form saves £400 000 in staff time

The UK Defence Ministry simplified a travel form used by over 750 000 people every year. The new form cut the error rate by half and also reduced processing time. It cost £12 000 to simplify, and saves £400 000 a year in staff time.

Case study 2

An internal report explaining the implications of a court ruling for a company's business

The headline and sub-text clearly state the conclusion; the body of the report backs it up.

The implications of *Kent* vs *Bank Five*:

We can be held responsible for giving clients bad investment advice!

Court rules against Bank Five

The Appeal Court has ordered Bank Five and one of its regional managers to pay $500 000 to a Mrs Kent after she received bad investment advice. The manager had advised her to place $300 000 in an investment scheme that was subsequently liquidated.

Court's decision centred on investment expertise

The court ruled that the manager's expertise should be judged against the higher standard expected of a major bank. This compares with the "lower" standard expected of an ordinary broker, who merely passes money on for investment without necessarily recommending that investment.

A court would have ruled against us too

A similar judgement would have been handed down for an intermediary working for our company. This is because our intermediaries are expected to advise clients not only on different products, but also on areas such as income tax, estate duty, risk and performance.

How do we ensure our intermediaries' expertise?

There are three ways of ensuring that our intermediaries have the appropriate skills and do not place themselves at risk:

- they must be CM-56 graduates, which should ensure that they give proper technical and legal advice
- they must do a proper needs analysis to show that any advice given was appropriate to the client's needs
- they must keep their advice in writing and hand a copy to the client; this makes it easier to resolve any subsequent disputes.

Case study 3

A report summarising what readers think of a corporate newsletter

Note how the headline and the first paragraph essentially tell the story. Afterwards, it's mere detail.

Readers give *Risk Newsletter* rave reviews

A definite thumb's up – that's what the September reader poll shows. Comments, varied from *The most succinct, informative and interesting publication of this category I have ever seen* to *What can I say?*. The 400 questionnaires returned represent a response rate of about 20% – quite high for a reader poll. We even had responses from Malta and Bermuda.

Content is a hit

89% rate the local briefs as good or excellent
91% rate the international briefs as good or excellent
94% rate the lead story as good or excellent

Typical comments: Topical, informative, pertinent, food for thought, hits the point hard, relates to issues of concern.

Format facilitates reading

93% read the entire newsletter
69% read it the day it arrives
66% refer it to other people

Typical comments: Succinct, punchy, brief but informative, nice and crisp, easy to read and pass on to colleagues.

Let's keep the hard-copy version

68% think an Internet-only newsletter is a bad idea
8% think an Internet-only newsletter is a good idea
51% think an Internet newsletter as an addition to the hard-copy version is worth considering

Typical comments: Not everyone's on the Internet; web version not as time-effective; we may forget to read it; hard copy is easier to circulate.

Case study 4

A stockbroking report advising caution on a popular share

Note the approach to the main headlines – they're on the side rather than in the body of the text.

Classy Stores: worth holding on to – but let's watch developments

Let's sell beyond 850c

The fair value of the share is around 800c–850c. If it moves beyond 850c, let's lighten our weighting.

Profits good, but slowing

Although *Classy Stores'* compound annual earnings growth over the past four years has been 77%, it has averaged only 12% over the past two years. Admittedly, that's off a high base and includes significant expansion costs, but it's an under-performance nonetheless.

Still worthy of a premium

Classy Stores' food side, which accounts for 35% of earnings, deserves a premium over most other food retailers and should probably be on a forward PE of 22.

The clothing side should also probably trade at a premium due to more stable earnings growth and stronger growth potential. It deserves a forward rating of 18.

Expansion is driving growth

In 2001, *Classy Stores* added 16% more floor space, significantly more than any competitor. That's expected to continue at around 10% annually until 2005. And most of this expansion is on the food side, where sales densities are three times those on the textile side.

Margins will be hurt by

Higher staff costs
This reflects additional staff taken on for previously outsourced food distribution and computer services.

Changing sales mix
As the sales mix shifts towards food sales, which tend to carry significantly lower mark-ups, margins will slip.

Case study 5

A report from an investment house about possible economic scenarios

Note how easy it is to take in the key points at a glance.

Three scenarios likely for global economy this year

We favour Scenario 1, which suggests slower OECD growth, looser monetary policies and great performances from bonds

Scenario 1 Status quo
Probability: 60%

Japan moves to avert depression
Japan eases economic and financial policies enough in the coming months to avert depression. Permanent income-tax cuts offset higher unemployment rates while a bridging plan for banks staves off a credit crunch.

Asia locked in recession
Tepid economic growth in Japan keeps monetary policy on the easy side and gradually pushes the yen-dollar exchange rate towards 170, locking Asia in recession.

US interest rates ease, supporting Western equity markets
Deflationary economic conditions in the East reinforce a modest slowdown in Western economies and lower OECD growth expectations. US interest rates ease, providing support for Western equity markets despite slower growth in corporate earnings.

Scenario 2 "Armageddon"
Probability: 30%

Run on Japanese banking system
Japan's economic and financial policies can't prevent a run on the Japanese banking system. Forced domestic selling of Japanese government bonds triggers a spike in bond yields to 2%, and the threat of capital flight causes the yen-dollar exchange rate to sink quickly to 190 by the end of the summer.

Emerging market crisis – and threats of recession in the West
Interest rates rise in emerging markets, leading to financial market crises in Russia, Brazil and China. Trade and capital-flow linkages between Japan and the emerging markets on the one hand, and the West on the other, threaten OECD nations with recession.

Bonds outperform equities
OECD countries cut interest rates, but too late to prevent a slump in corporate earnings growth. Global bonds end up outperforming global equities by a wide margin.

Scenario 3 Eastern surprise
Probability: 10%

Macro-economic recovery in the East
Asia greatly eases monetary policy and Japan slashes taxes, spurring hopes for a macro-economic recovery in the Asia-Pacific region.

Corporate profits improve
Financial and corporate restructuring activity in the area improves corporate profit expectations. This is further boosted by the creation of an international secondary market for Asia's non-performing debt.

Asian, Japanese equities outperform global averages
Asian and Japanese equities outperform global averages, successful reflation efforts in the East raise inflation fears in the West and OECD bond prices fall sharply.

Case study 6

One of the entries for a competition for the best new building of the year

Note the use of a summary even though the detail is not that extensive. It will help the judges remember the main points as they probably have a pile of other entries to read.

This is a good example of how even a long report can be easy to digest when it is written using the techniques you have just learned.

How *Conquest House* satisfies the judging criteria

Conquest Bank's entry for the 2002
Building of the Year competition

Contents

How Conquest House satisfies the judging criteria

Summary

1 Overall impact – The most visible landmark in Sandton

With its extensive glass frontage, polished granite façade and distinctive water features, this 18-storey building is arguably the most visible landmark in Sandton. It makes a powerful statement of style, technology and commerce.

2 Functionality – Not just a pretty face

Over 800 bankers ply their complex trade in Conquest House, thanks to a productivity-enhancing blend of high technology and common-sense design. From its catering and conference facilities to its fully-equipped gym and ample visitors' parking, the building is not just a pretty face.

3 Financial viability – The numbers of a commercial venture

It may be a corporate head office, but Conquest House's numbers read more like those of a hard-nosed commercial venture. It was completed in only 18 months thanks to diligent pre-planning, innovative construction and a 100% safety record. The ratio of gross lettable area to gross building area is a high 95%. The rentals, reflected in an initial yield of 9.8%, are below market rates.

4 Architecture – A blend of local and international styles

Its inspiration is from Singapore and the US, but its final expression is unmistakably local – an ultra-modern skyscraper set in a stark and moody African landscape. An earth-coloured piazza with rustic steps and waterfalls rises to a solid, polished granite façade which finally sweeps skywards into an impressive glass superstructure.

5 Innovative features – State-of-the-art internal systems

The building breaks new ground in innovation, boasting such novelties as "intelligent" computerised lifts; a centralised air-conditioning system which allows individual temperature setting; centrally controlled electrical, data and telecommunications networks; heat-reflective glass; and extensive under-cover parking for users and visitors alike.

Overall impact

The most visible landmark in Sandton

Can be seen from Midrand

With its extensive glass frontage, polished granite façade and distinctive water features, this 18-storey building is arguably the most visible landmark in Sandton. This is all the more so as it is situated on the highest geographical point of the area and, on a clear day, can be seen from as far away as Midrand.

A symbol of Sandton

Conquest House makes a powerful statement of style (through its ultra-modern design), technology (through the use of state-of-the-art interior equipment) and commerce (by being the focal point of the greatest concentration of banks in Johannesburg). It is as much a symbol of Sandton today as Sandton Towers used to be when it was the only skyscraper in the area.

Home to six major banks

Conquest House is now inseparable from the greater Conquest Place which surrounds it. This 80 000 m^2 complex – nearly 11% of the total lettable area in the Sandton business district – is home to six local and international merchant banks.

Functionality

Not just a pretty face

Around-the-clock activity

Conquest House is home to over 800 of the country's most skilled bankers. They ply their complex trade around the clock on eighteen levels of productively utilised floor space. Key departments and group companies – previously scattered around Sandton in different buildings – now have direct access to each other and are able to maximise synergies.

Maximum productivity

The building blends high technology with common-sense design to create a working environment that ensures maximum productivity. In addition, there are full catering facilities, an ultra-modern conference facility, a fully-equipped gym, ample and secure visitors' parking as well as nearby shops and restaurants. The building is a hive of activity in which a lot of work – and relaxation – gets done.

Long-term growth potential

All aspects of the building, from parking to outside terraces, flow naturally into the greater Conquest Place complex. This gives the building long-term "extendability" and growth potential.

Financial viability

The numbers of a commercial venture

Completed in only 18 months

Conquest House's numbers read more like those of a hard-nosed commercial venture than a corporate head office. The fact that it was completed in only 18 months (which included two quiet Decembers) spelled major cost savings.

A good safety record

The speed of construction was due mainly to:

- diligent pre-planning (everything from building materials to services needed were ordered well in advance)
- innovative construction techniques (the core and the sides went up rapidly enough to meet the deadline, yet carefully enough to ensure quality)
- a 100% safety record (impressive for a tall, rapidly constructed building).

Below-market rental rates

The short construction time, the use of extensive parkade parking (which is less expensive than basement parking) and the high design efficiency (gross lettable area is 95% of gross building area) resulted in final rentals well below market rates. The initial yield was a competitive 9.8%.

Architecture

A blend of local and international styles

Overseas inspiration, local execution

Conquest House got its inspiration from the high-tech financial centres of Singapore and the US, but its final expression in terms of aesthetics and architecture is unmistakably local. Only in South Africa could such a contrast work – an ultra-modern skyscraper planted in a landscape that is uniquely African.

A contrast from base to summit

The building is securely anchored in an African, earth-coloured piazza setting, complete with rustic steps, peaceful waterfalls and hard boulders – all softened with azaleas and other local shubbery. The building then rises tentatively, displaying a solid base of polished, tan-coloured granite. It then sweeps skywards into 5 000m^2 of brilliant, diamond-like glass which reflects the mood of the sky – from bright blue at midday to golden orange at sunset.

Décor mirrors exterior theme

The building's contrasting setting finds expression inside too, where classy, modern furnishings and high-tech fittings co-exist harmoniously with warm ethnic colours, black granite counters and natural wood panelling.

Innovative features

State-of-the-art internal systems

Intelligent lifts

There are no floor buttons inside the "intelligent" computerised *Miconic 10* lifts. Instead, passengers select their floors outside the lifts and are then routed to their destinations in appropriate numbers and at appropriate intervals. The result is less waiting time.

A centralised data network

The electrical, data and telecommunications networks – the most advanced of their kind – are centrally controlled, yet offer great flexibility: staff can move offices without having to change telephone or computer extensions. Over 200km of fibre optic cable allow multiple levels of security.

Individual office temperature settings

Finally, the silver metallic coating and blue-tinted interlayer on the heat-reflective glass exterior eliminates 76% of radiant solar energy before it enters the building, significantly reducing air-conditioning costs. The air-conditioning system itself offers individual office temperature settings while simultaneously optimising air flow and heating throughout the building.

Bibliography

Better Business Writing
MARYANN V. PIOTROWSKI

ISBN 0-7499-1157-3
Judy Piatkus
5 Windmill Street
London W1
England

Business Writing for Busy People
PHILIP R. THEIBERT

ISBN 1-56414-223-X
Career Press
3 Tice Road
PO Box 687
Franklin Lakes, NJ 07417
USA

Clarity for Lawyers
The use of plain legal language in legal writing
MARK ADLER

ISBN 1-85328-087-9
The Law Society
113 Chancery Lane
London WC2A 1PL

Communicating, or Just Making Pretty Shapes
COLIN WHEILDON
Newspaper Advertising Bureau of Australia
1st floor 77 Berry Street
North Sydney NSW 2060
Australia

Conquer the Information Mountain
A guide to eliminating paper and electronic junk in the office
DECLAN TREACY

ISBN 0-09-927095-1
Arrow Books
20 Vauxhall Bridge Road
London SW1V 2SA
England

Corporate Speak
The use of language in business
FIONA CZERNIAWSKA

ISBN 0-333-67477-4
MacMillan Press
Houndmills
Basingstoke
Hampshire RG21 6XS
England

Cover Letters
MICHELLE TULLIER, PhD.

ISBN 0-679-77873-X
Princeton Review Publishing, LLC
2315 Broadway, 2nd floor
New York, NY 10024
USA

Dynamics in Document Design
KAREN A. SCHRIVER

ISBN 0-471-30636-3
John Wiley & Sons, Inc
605 Third Avenue
New York, NY 10158
USA

Executive's Book of Quotations
A guide to the right quote for every occasion
JULIA VITULLO-MARTIN and J. ROBERT MOSKIN

ISBN 0-19-507836-5
Oxford University Press, Inc
200 Madison Avenue
New York, NY 10016
USA

Gobbledygook
An anthology of bad official writing collected by Plain English Campaign
PLAIN ENGLISH CAMPAIGN

ISBN 0-04-827107-1
www.plainenglish.co.uk

How to Get Your Message Across
A practical guide to power communication
DR DAVID LEWIS

ISBN 0-285-63348-1
Souvenir Press
43 Great Russell Street
London WC1B 3PA
England

How to Take the Fog Out of Business Writing
ROBERT GUNNING and RICHARD A. KALLAN

ISBN 0-85013-232-0
Dartnell Corporation
4660 N Ravenswood Avenue
Chicago, IL 60640-4595
USA

Investor-friendly Annual Reports
ROBERT GENTLE

ISBN 0-620-25355-X
Plain Business Writing
PO Box 785553
Sandton 2146
South Africa

Language on Trial
The plain English guide to legal writing
PLAIN ENGLISH CAMPAIGN

ISBN 1-86105-006-2
www.plainenglish.co.uk

Making a Good Layout
LORI SIEBERT and LISA BALLARD

ISBN 0-89134-423-3
North Light Books
(an imprint of F&W Publications)
1507 Dana Avenue
Cincinnati
Ohio 45207
USA

Mastering Business Communication
L.A. WOOLCOTT and W.R. UNWIN

ISBN 033-333-5295

Maverick
The success behind the world's most unusual workplace
RICARDO SEMLER

ISBN 0-09-932941-7
Arrow Books
20 Vauxhall Bridge Road
London SW1V 2SA
England

McCormack on Communicating
MARK H. McCORMACK

ISBN 0-7126-7503-5
Century
Random House
20 Vauxhall Bridge Road
London SW1V 2SA
England

NTC's Business Writer's Handbook
Business communication from A to Z
ARTHUR H. BELL

ISBN 0-8442-5913-6
NTC Publishing Group
4255 West Touhy Avenue
Lincolnwoold (Chicago)
Illinois 60646-1975
USA

Ogilvy on Advertising
DAVID OGILVY

ISBN 1-85375-196-0
Prion Books
Imperial Works
Perren Street
London NW5 3ED
England

On Writing Well
WILLIAM K. ZINSSER

ISBN 00-627352-33
Harper Reference

Plain English at Work
A guide to business writing and speaking
EDWARD P. BAILEY

ISBN 0-19-510449-8
Oxford University Press, Inc
198 Madison Avenue
New York, NY 10016
USA

Plain English for Lawyers
Fourth edition
RICHARD C. WYDICK

ISBN 0-89089-994-0
Carolina Academic Press
700 Kent Street
Durham
North Carolina 27701
USA

Plain English Pilot Program
Selected plain English samples
SECURITIES AND EXCHANGE COMMISSION (USA)

ISBN 1-886100-11-X
Published by Bowne
www.bowne.com

For more information, visit the Securities and Exchange Commission's website at *www.sec.gov*

The Complete Plain Words
SIR ERNEST GOWERS

ISBN 0-405119-97
Penguin Group
27 Wrights Lane
London W8 5TZ

The Concise Oxford Dictionary of Quotations

ISBN 0-928007-01
Oxford Paperbacks
www.oup.co.uk

The Elements of Style
WILLIAM STRUNK JR and E.B. WHITE
ISBN 0-205-19158-4
Allyn & Bacon
A Simon & Schuster Company
Needham Heights
Massachusetts 02194
USA

The Language Instinct
STEVEN PINKER
ISBN 0-14-017529-6
Penguin Books
27 Wrights Lane
London W8 5TZ
England

The Plain English Approach to Business Writing
Revised edition
EDWARD P. BAILEY
ISBN 0-19-511565-1
Oxford University Press, Inc
198 Madison Avenue
New York, NY 10016
USA

The Plain English Guide
MARTIN CUTTS
ISBN 0-986624-32
Oxford University Press
www.oup.co.uk

The Plain English Story
PLAIN ENGLISH CAMPAIGN
www.plainenglish.co.uk

The Power of Simplicity
JACK TROUT with STEVE RIVKIN

ISBN 0-07-065362-3
McGraw-Hill, Inc
2 Penn Plaza
New York
NY 10121
USA
www.mcgraw-hill.com

Utter Drivel
A decade of jargon and gobbledygook as recorded by Plain English Campaign
PLAIN ENGLISH CAMPAIGN

ISBN 0-86051-949-X
www.plainenglish.co.uk

Verbiage for the Verbose
Unravel over 250 fun word challenges
PETER GORDON

ISBN 0-8362-5193-8
Andrews McMeel Publishing
4520 Main Street
Kansas City
Missouri 64111
USA

Why We Buy
The science of shopping
PACO UNDERHILL

ISBN 0-684-84913-5
Simon & Schuster
Rockefeller Center
1230 Avenue of the Americas
New York, NY 10020
USA

Write Language
ALLAN PEASE and PAUL DUNN

ISBN 0-9593658-3-4
Camel Publishing Company
50 Hilltop Road
Clareville Beach
NSW 2107
Australia

Write Tight
How to keep your prose sharp, focused and concise
WILLIAM BROHAUGH

ISBN 0-89879-548-6
Writer's Digest Books
Cincinnati, Ohio
USA

Notes

1. **Plain Language Prospectuses**
   What we can learn from the US
   ROBERT GENTLE ©1998

2. **The Scribes Journal of Legal Writing (Volume 6, 1996–97)**
   Writing for dollars, writing to please
   JOSEPH KIMBLE

   Thomas Cooley Law School
   Lansing, Michigan
   USA
   email: kimblej@cooley.edu

3. **The Plain English Campaign (UK)**
   www.plainenglish.co.uk